REVISE PEARSON EDEXCEL G

Mathematics
Higher

TEN-MINUTE TESTS

Series Consultant: Harry Smith

Author: Ian Bettison and Su Nicholson

Published by Pearson Education Limited, 80 Strand, London, WC2R 0RL.

www.pearsonschoolsandfecolleges.co.uk

Copies of official specifications for all Pearson qualifications may be found on the website: qualifications.pearson.com

Text and illustrations © Pearson Education Ltd 2019
Typeset and illustrated by Newgen KnowledgeWorks Pvt. Ltd., Chennai, India
Produced by Newgen Publishing UK
Cover illustration by Miriam Sturdee

The rights of Ian Bettison and Su Nicholson to be identified as authors of this work has been asserted by them in accordance with the Copyright, Designs and Patents Act 1988.

First published 2019

22 21 20 19
10 9 8 7 6 5 4 3 2 1

British Library Cataloguing in Publication Data
A catalogue record for this book is available from the British Library

ISBN 978 1 2922 9430 8

Printed in Italy by L.E.G.O. SpA

Notes from the publisher
1. While the publishers have made every attempt to ensure that advice on the qualification and its assessment is accurate, the official specification and associated assessment guidance materials are the only authoritative source of information and should always be referred to for definitive guidance.
Pearson examiners have not contributed to any sections in this resource relevant to examination papers for which they have responsibility.
2. Pearson has robust editorial processes, including answer and fact checks, to ensure the accuracy of the content in this publication, and every effort is made to ensure this publication is free of errors. We are, however, only human, and occasionally errors do occur. Pearson is not liable for any misunderstandings that arise as a result of errors in this publication, but it is our priority to ensure that the content is accurate. If you spot an error, please do contact us at resourcescorrections@pearson.com so we can make sure it is corrected.

How to use this book

This book is designed to help you test yourself on the knowledge and skills you will need for your GCSE (9–1) Mathematics exam. The book contains 44 short tests, covering the whole of your specification and you should spend 10 minutes on each test.

Easy-to-use answers with hints, marking tips and working will allow you to mark your tests quickly and accurately. This will help to boost your confidence in areas where you get a high mark. It will also help you to plan your revision effectively by focusing on the areas in which you get a lower mark and could improve with further practice. Some extra pointers are given at the end of each test to help you develop your personal revision plan.

If you have the Revise Pearson Edexcel GCSE (9–1) Mathematics Revision Guide you can follow the links at the top of each test for more help with that topic.

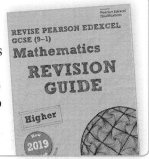

1 Look out for these **core skill** questions – these skills and topics come up year after year so make sure you are confident with them.

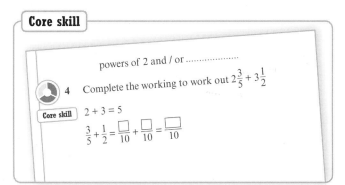

2 Look at this **scale** next to each question in a test. It will tell you how difficult the question is.

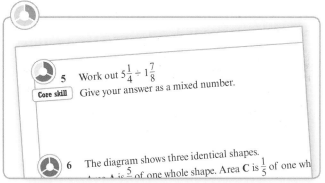

3 Use the **QR code** to jump straight to the answers, or go to the back of the book. Use the **hints** in the answers to help you understand any questions you answered incorrectly.

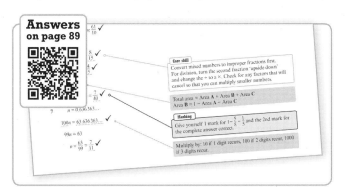

4 Record your mark for each question in the box next to the question. Then total your number of marks for the whole test at the bottom of the right-hand page.

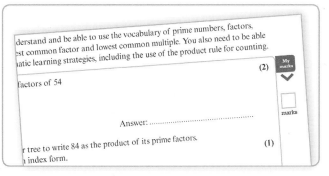

5 Use your score to help you **plan** your revision. If you need more help, look at the Revise Pearson Edexcel GCSE (9–1) Mathematics Revision Guide.

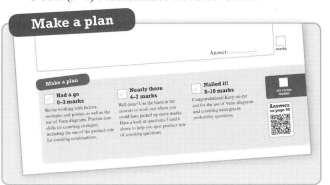

6 Use the contents page to **record** your test scores. Create your own revision plan by ticking the boxes to show your confidence level with every topic across the whole specification.

Contents

A small bit of small print

Pearson Edexcel publishes Sample Assessment Material and the Specification on its website. This is the official content and this book should be used in conjunction with it. The questions have been written to help you revise topics and practise the skills you will need in your exam.

Remember – the real exam questions may not look like this.

1 Fractions and decimals

  10 You need to be able to work with fractions and mixed numbers, and be able to calculate terminating and recurring decimals **without** using a calculator.

My marks

1 What decimal is equivalent to $\frac{9}{20}$?
Tick **one** box. **(1)**

☐ **A** 0.045
☐ **B** 0.036
☐ **C** 0.36
☑ **D** 0.45

marks

2 Which set of values is in the correct order of size, from smallest to largest?
Tick **one** box. **(1)**

☐ **A** $61\%, \frac{2}{3}, \frac{13}{20}, 0.67$

☑ **B** $\frac{2}{3}, 61\%, \frac{13}{20}, 0.67$

☐ **C** $\frac{13}{20}, 61\%, \frac{2}{3}, 0.67$

☐ **D** $61\%, \frac{13}{20}, \frac{2}{3}, 0.67$

marks

3 **(a)** Which of these fractions can be written as a terminating decimal?
Tick **one** box. **(1)**

☐ **A** $\frac{7}{24}$

☐ **B** $\frac{11}{65}$

☐ **C** $\frac{9}{40}$

☐ **D** $\frac{5}{28}$

marks

(b) Fill in the gaps for the following statement. **(1)**

A fraction can be written as a terminating decimal if the prime factors of the

.... of the fraction only consist of

powers of 2 and / or ....

marks

4 Complete the working to work out $2\frac{3}{5} + 3\frac{1}{2}$ **(2)**

 Core skill $2 + 3 = 5$

$$\frac{3}{5} + \frac{1}{2} = \frac{\square}{10} + \frac{\square}{10} = \frac{\square}{10}$$

Answer: ....

marks

5 Work out $5\frac{1}{4} \div 1\frac{7}{8}$

Core skill   Give your answer as a mixed number. (2)

Answer: ....

marks

6 The diagram shows three identical shapes.
Area **A** is $\frac{5}{8}$ of one whole shape. Area **C** is $\frac{1}{5}$ of one whole shape.
Work out Area **B** as a fraction of one whole shape. (2)

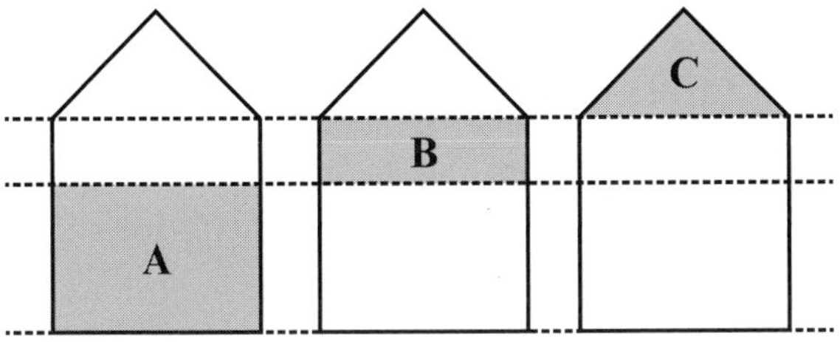

Answer: ....   marks

7 Work out the recurring decimal $0.6\dot{3}$ as a fraction in its simplest form.
You must show your working. (2)

Let $n = 0.636363...$

...................$n =$ ....

Answer: ....   marks

Make a plan

✓ **Had a go** 0–4 marks	✓ **Nearly there** 5–8 marks	✓ **Nailed it!** 9–12 marks
Recap the operations on fractions and decimals. Focus on the core skills of finding a common denominator to add and subtract fractions and on the methods for multiplication and division.	Well done! Use the hints in the answers to work out where you could have picked up more marks. Remember that if a question asks you to give your answer in its simplest form, you need to identify common factors in the numerator and denominator. You also need to learn how to write recurring decimals as fractions.	Congratulations! Keep an eye out for use of fractions in diagrams as well – you might need to pick which operation is appropriate in problem-solving questions.

MY TOTAL
MARKS

Answers
on page 89

2 Factors, multiples and counting

  You need to understand and be able to use the vocabulary of prime numbers, factors, multiples, highest common factor and lowest common multiple. You also need to be able to apply systematic learning strategies, including the use of the product rule for counting.

 My marks

1 Write down all the factors of 54 **(2)**

Answer: ....

marks

2 Complete the factor tree to write 84 as the product of its prime factors.
Give your answer in index form. **(1)**

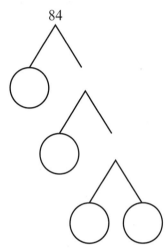

84

Answer: 84 = ...

marks

3 Use your answer to Question 2 and complete the working and Venn diagram below to help you work out the highest common factor (HCF) of 84 and 180 **(3)**

As a product of its prime factors 180 = ...

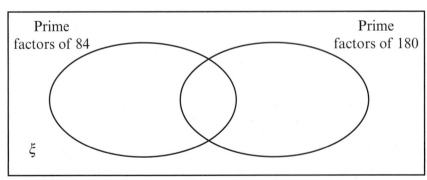

Prime factors of 84

Prime factors of 180

ξ

Answer: HCF = ...

marks

4 Use your answer to Question 3 to help you work out the lowest common multiple (LCM) of 84 and 180 **(1)**

Answer: LCM = ...

marks

5 Work out how many 4-digit numbers could be made using these number cards. **(1)**

Core skill

| 5 | | 1 | | 7 | | 8 |

Answer: ....

marks

6 A password is made up of three characters. The first character is a letter. The second character is a non-zero even number. The third character is an odd number. Work out the total number of possible passwords. **(2)**

Total number of possible passwords = ☐ × ☐ × ☐

Answer: ....

marks

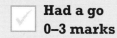
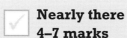
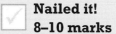

Revision Guide pages 2, 3

3 Powers and roots

 You need to be able to use positive integer powers and associated real roots, as well as recognise powers of 2, 3, 4 and 5. You also need to be able to estimate powers and roots of any given positive number. In addition, you need to be able to calculate with roots and with integer and fractional indices.

1 What is 32 as a power of 2?
Tick **one** box. (1)

☒ **A** 2^4

☑ **B** 2^5

☐ **C** 2^6

☐ **D** 2^7

✓

1 marks

2 Work out

(a) $\sqrt{225}$ (1)

Answer: ....15.... ✓

1 marks

(b) 5^3 (1)

Answer: ...125... ✓

1 marks

3 Write as a single power of 3

(a) $(3^2)^3 \times 3^0$ (1)

Answer: ...3^6... ✓

1 marks

(b) $\dfrac{3^4 \times 3^2}{3}$ (1)

Answer: ...3^5 ✓ ...

1 marks

4 Which of these is the closest estimate to $\sqrt{180}$?
Tick **one** box.

☐ **A** 11.8
☑ **B** 13.4 ✓
☐ **C** 12.9
☐ **D** 15.1

marks: 1

 5 Given $x^{-\frac{3}{2}} = \dfrac{27}{125}$, complete the working and find the exact value of x. (3)

Core skill

$$x^{-\frac{3}{2}} = \frac{27}{125}$$

$$x^{\frac{3}{2}} = \frac{\boxed{125}}{\boxed{27}} \checkmark$$

$\left(\dfrac{125}{27}\right)^{1/3} = \dfrac{5}{3}$

$\left(\dfrac{5}{3}\right)^2 = \dfrac{25}{9}$

Answer: ....

marks: 1

 6 Find the value of x when $2^{3x-5} = \dfrac{1}{8}$ (3)

Answer: ....

marks: 0

Make a plan

☑ **Had a go** 0–4 marks	☑ **Nearly there** 5–8 marks	☑ **Nailed it!** 9–12 marks
Recap your work on indices to help with this topic. Make sure you know the square numbers up to 15^2 and their corresponding square roots without a calculator.	Well done! Make sure you learn the laws of indices and know how to apply them.	Congratulations! Keep an eye out for the use of indices in algebra questions as well.

MY TOTAL MARKS

7

Answers on page 91

4 Exact answers and standard form

  You need to be able to calculate exactly with fractions, surds and multiples of π. In addition, you should be able to simplify surd expressions involving squares and rationalise the denominator. You also need to be able to calculate with and interpret numbers given in standard form $A \times 10^n$, where $1 \leqslant A < 10$ and n is an integer.

 1 Simplify $\sqrt{75}$ **(1)**

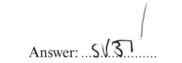 $\sqrt{25}^1 \times \sqrt{3}^1$ $5\sqrt{3}^1$

Answer: $5\sqrt{3}$ marks

 2 Rationalise the denominator of $\dfrac{7}{2\sqrt{3}}$ **(1)**

Answer: ....   marks

 3 $\sqrt{72} + \sqrt{242} = k\sqrt{2}$ where k is an integer.
Find the value of k.
You must show your working. **(2)**

$\sqrt{72} + \sqrt{242} = \sqrt{\rule{2cm}{0.4pt}} + \sqrt{\rule{2cm}{0.4pt}}$

Answer: ....   marks

 4 Work out $(6.7 \times 10^6) + (3.4 \times 10^5)$ giving your answer in standard form.
You must show your working. **(2)**

Core skill

$6.7 \times 10^6 = $ ....

$3.4 \times 10^5 = $ ....

Answer: ....   marks

 5 Work out $(1.4 \times 10^{-5}) \div (2 \times 10^{-2})$ giving your answer in standard form.
You must show your working. **(2)**

Core skill

$(1.4 \times 10^{-5}) \div (2 \times 10^{-2}) = (1.4 \div 2) \times ($...................$)$

Answer: ....   marks

My marks

6 The perimeter of a circle is $\frac{12}{5}\pi$ cm. The perimeter of a larger circle is $\frac{10}{3}\pi$ cm.

How much longer is the perimeter of the larger circle.

Give your answer in terms of π. You must show your working. **(2)**

$$\frac{10}{3}\pi - \frac{12}{5}\pi = \left(\frac{\boxed{} - \boxed{}}{15}\right)\pi$$

Answer: .... cm

 marks

 7 The area of circle **A** is $\frac{16}{9}\pi\,\text{cm}^2$.

The area of circle **B** is $36\pi\,\text{cm}^2$.

How many times greater is the area of circle **B** than the area of circle **A**?

Give your answer as a fraction.

You must show your working. **(2)**

Answer: ....

marks

Make a plan

 Had a go
0–4 marks

Make sure you can carry out calculations on standard form numbers without a calculator. Practise working with surds and recap calculating with fractions.

 Nearly there
5–8 marks

Well done! Practise spotting factors of numbers which are square numbers to help you simplify surds quickly without a calculator.

 Nailed it!
9–12 marks

Congratulations! Keep an eye out for the use of standard form in measure questions involving very large or very small numbers. Remember that geometry questions that ask for exact answers may need to be expressed in terms of surds and π.

MY TOTAL MARKS

Answers on page 91

5 Calculator skills

 You need to be able to use your calculator to calculate with roots and with integer and fractional indices. Make sure you know how to use all the functions on your calculator.

1 Write 7.6×10^{-4} as an ordinary number. **(1)**

Answer: 0.00076 ✓

| |
| marks |

2 Write 580 000 000 as a number in standard form. **(1)**

Answer: 5.8×10^8 ✓

| |
| marks |

3 Find the value of $(3.6 - 0.55)^2 + \sqrt[3]{10.648}$
You must show your working. **(2)**

$(3.6 - 0.55)^2 =$ 9.3025

$\sqrt[3]{10.648} =$ 2.2

✓✓

Answer: 11.5025

(11.5)

| |
| marks |

4 Find the value of $\dfrac{\sqrt{12.5 + 3.4}}{4.2^3}$
Write down all the figures on your calculator display.
You must show your working. **(2)**

$\sqrt{12.5 + 3.4} =$ 3.987480407

$4.2^3 =$ 74.088

✓

Answer: 0.05352086718 ✓

(0.05)

| |
| marks |

5 Work out $(3.72 \times 10^{-4}) \times (2.1 \times 10^7)$
Give your answer in standard form. **(2)**

7812

✓✓

Answer: 7.812×10^3

| |
| marks |

6 Work out the value of $\dfrac{2.625 \times 10^5}{5.25 \times 10^{-3}}$

Core skill Give your answer in standard form. (2)

Answer: .... marks

 7 In 2019, the population of the UK was estimated as 6.69×10^7
The area of the UK is $2.425 \times 10^5 \, km^2$.
Calculate the mean number of people per square kilometre in the UK.
Give your answer to an appropriate degree of accuracy. (2)

Answer: .... marks

Make a plan

☑ **Had a go**
0–4 marks

Make sure you know how to enter fractions, standard form numbers and powers on your calculator. Always practise with the same calculator you will use in your exam.

☑ **Nearly there**
5–8 marks

Well done! Use the hints in the answers to work out where you could have picked up more marks. Make sure you practise your calculator skills, and check that your answers make sense.

☑ **Nailed it!**
9–12 marks

Congratulations! Keep an eye out for the use of standard form in measure questions involving very large or very small numbers.

 8
MY TOTAL MARKS

Answers on page 92

6 Estimation and accuracy

You need to be able to round numbers to an appropriate degree of accuracy and estimate answers using approximation. You also need to be able to apply and interpret limits of accuracy, including upper and lower bounds and simple error intervals.

1 By rounding each value to 1 significant figure, work out an **estimate** for $\dfrac{5.23 \times 3.47}{0.472}$
You must show your working.

$5.23 \approx$ ....**5**....

$3.47 \approx$ ....**3**....

$0.472 \approx$ ....**0.5**....

$\frac{15}{0.5}$

Answer: ....**= 30** ✓.........

My marks

2 marks

2 The dimensions of a rectangle are 18 cm and 11 cm, correct to 2 significant figures.

(a) Calculate the lower bound for the perimeter of the rectangle.
You must show your working.

Lower bound for 18 cm = 「**17.5**」 cm
Lower bound for 11 cm = 「**10.5**」 cm
Lower bound for perimeter = ...~~17.5 or 28~~....

$17.5 + 10.5 = 28$)
$28 \times 2 = 56$ ✓

Answer: ....**56**.... cm

2 marks

(b) Calculate the lower bound for the area.

$17.5 \times 10.5 =$

Answer: ..**183.75** ✓.. cm²

1 marks

3 The area, A, of a rectangle is 26 cm² correct to 2 significant figures.
The length, l, of the rectangle is 5.1 cm correct to 2 significant figures.
Calculate the upper bound for the width, w, of the rectangle.
You must show your working.

	Lower bound	Upper bound
Area of rectangle	25.5	26.5
Length of rectangle	5.05	5.15

Upper bound for width =

$w = \dfrac{area}{length} = \dfrac{26.5}{5.15} =$

Answer: **5.146** cm

3 marks

11

4 The curved surface area of a cylinder is 120 cm², to the nearest cm².
The radius is 3.1 cm, correct to 1 decimal place.

 Core skill

Find the height of the cylinder to an appropriate degree of accuracy.
You must show your working.

(3)

	Lower bound	Upper bound
Surface area	119·5	120·5 ✓✓
Radius	3·05	3·15

Upper bound for height = ...$\dfrac{120.5}{2 \times \pi \times 3.05}$........

Lower bound for height = ...$\dfrac{119.5}{2 \times \pi \times 3.05}$..........

Answer: ....

1 marks

5 Ellie uses a calculator to find the value of a number n. She only writes down the first two
digits of the answer on her calculator display. She writes down 5.3
Write down the error interval for the actual value of n.

(1)

$5.25 \quad < 5.3 < 5.35$ ✗

$5.3 \searrow 54$

Answer: ....

0 marks

---

7 Brackets and factorising

 You need to be able to expand and factorise different types of algebraic expression, including quadratic expressions and expressions with three linear factors.

1 Fill in the gaps to complete the factorisation. **(2)**

$4x^2 - 6x = $ ....**4**..... (....**x**.......... − ...**2c**...........)

2 Complete the working to expand and simplify $3(x + 7) + 5(2x - 3)$ **(2)**

$3(x + 7) + 5(2x - 3) = $ ....**3**......$x + 21 + 10x - $.**15**.............. ✓

$13x + 6$ ✓

Answer: ....

 0 marks

 2 marks

3 Expand and simplify $(x + 4)(x + 5)$ **(1)**

$(x + 4)(x + 5)$

$x^2 + 5x + 4x + 20$
$x^2 + 9x + 20$ ✓

Answer: ....

 1 marks

4 Work out the correct factorisation of $x^2 - 5x - 14$
Tick **one** box. **(1)**

Core skill

✓ **A** $(x + 7)(x - 2)$ ✓
☐ **B** $(x - 7)(x + 2)$
☐ **C** $(x - 2)(x - 7)$
☐ **D** $(x + 2)(x + 7)$

 1 marks

5 Fill in the gaps to complete the factorisation. **(1)**

$x^2 - 64 = (x$ ....$)(x$ ....$)$

marks

6 Expand and simplify $(2x - 3)(3x + 4)$ **(1)**

Answer: ....

marks

7 Fill in the gaps to complete this expansion. (2)

$$(x - 2)(x + 4)(x + 3) = (x^2 + \text{.............................})(x + 3)$$

$$= x^3 + \text{.............................}$$

marks

 8 Work out the correct factorisation of the following expression.
$2x^2 + 5x - 3$

Core skill Tick **one** box. (1)

- ☐ **A** $(2x - 1)(x + 3)$
- ☐ **B** $(2x + 3)(x - 1)$
- ☐ **C** $(2x + 1)(x - 3)$
- ☐ **D** $(2x - 3)(x + 1)$

marks

 9 Work out the correct factorisation of the following expression.
$12x^2 - x - 6$

Core skill Tick **one** box. (1)

- ☐ **A** $(4x + 3)(3x - 2)$
- ☐ **B** $(6x + 2)(2x - 3)$
- ☐ **C** $(6x - 2)(2x + 3)$
- ☐ **D** $(3x + 2)(4x - 3)$

marks

Make a plan

 Had a go
0–4 marks
Try revising the expansion of simple brackets. Make sure you multiply all the parts of all the brackets. Factorising quadratic expressions is a core skill so it's worth taking the time to revise it properly.

 Nearly there
5–9 marks
Well done! Use the hints in the answers to work out where you could have picked up more marks. Remember that if a question asks you to factorise a quadratic, you can check your factorisation afterwards by expanding your answer.

Nailed it!
10–12 marks
Congratulations! Keep an eye out for more complicated expansions. If one bracket is squared it can sometimes help to write it out twice.

MY TOTAL MARKS

Answers on page 93

8 Algebraic manipulation

  You need to be able to substitute into formulae, change the subject of a formula and simplify expressions involving indices and algebraic fractions.

 1 The formula for converting from degrees Celsius to degrees Fahrenheit is

$$F = \frac{9}{5}C + 32$$

Calculate C when $F = 482$ **(1)**

Answer: .... °C

marks

 2 $E = mc^2$

Make c the subject of the formula. **(2)**

Core skill

Answer: $c =$ ....

marks

 3 $A = 2b - c$

Which of these is **not** a correct rearrangement of the above formula?

Tick **one** box. **(1)**

☐ **A** $c = 2b - A$

☐ **B** $b = \frac{1}{2}(A + c)$

☐ **C** $b = \frac{c + A}{2}$

☐ **D** $b = \frac{A - c}{2}$

marks

 4 Simplify $\dfrac{5x^7y^2 \times 6x^6y^{-1}}{3x^4y^{-2}}$ **(2)**

Answer: ....

marks

 5 $D = \dfrac{2(ab - a^2 + 3c)}{b^2 - c^2}$

Calculate the value of D when $a = 2.1$, $b = 3.7$ and $c = 1.6$

Give your answer to 3 significant figures. **(1)**

Answer: ....

marks

6 Fill in the gaps to make x the subject. (2)

$$a(x - y) = 4(b - x)$$

$$ax\text{............} = 4b\text{............}$$

$$ax\text{............} = 4b + \text{............}$$

$$x(\text{...............}) = \text{...............}$$

$$x = \frac{\text{...............}}{\text{...............}}$$

Answer: ....   marks

 7 $\dfrac{x^2 - 3x + 2}{x^2 + 2x - 8}$ is simplified.

Which of these is correct?
Tick **one** box. (1)

☐ **A** $\dfrac{-3x + 2}{2x - 8}$

☐ **B** $\dfrac{-x + 2}{-8}$

☐ **C** $\dfrac{x - 1}{x + 4}$

☐ **D** $\dfrac{x}{4}$

marks

 8 Simplify fully $\dfrac{b + 2}{12b} + \dfrac{5}{8b}$ (2)

Answer: ....   marks

9 Functions and proof

  You need to know the difference between an equation and an identity, use algebra to construct mathematical proofs and find composite and inverse functions.

1 Choose the correct word from the box to complete this sentence. **(1)**

| identity | equation | expression |

$3(x + 2) = 3x + 6$ is an example of an ....

marks

2 An odd number is written algebraically as $2n + 1$
What is the correct expression for the next odd number?
Tick **one** box. **(1)**

☐ **A** $2n + 3$
☐ **B** $2n + 2$
☐ **C** $3n + 1$
☐ **D** $n + 2$

 marks

3 The function f is such that $f(x) = 2x + 7$

 (a) Find $f(3)$ **(1)**

Answer: ....

 marks

(b) Solve $f(x) = 17$ **(1)**

Answer: ....

 marks

4 $f(x) = 3x - 1$
$g(x) = x^3$
Calculate the value of $fg(2)$
Tick **one** box. **(1)**

☐ **A** 125
☐ **B** 23
☐ **C** 17
☐ **D** 25

 marks

5 $f(x) = 5x + 3$
Find the inverse function f^{-1}
 Tick **one** box. **(1)**

☐ **A** $f^{-1}(x) = 5x + 3$
☐ **B** $f^{-1}(x) = \dfrac{x + 3}{5}$
☐ **C** $f^{-1}(x) = 5x - 3$
☐ **D** $f^{-1}(x) = \dfrac{x - 3}{5}$

 marks

6 Fill in the gaps to complete the mathematical argument to show that the sum of three consecutive even numbers is a multiple of 6 **(3)**

$2n +$ .... $+ (2n + 4) = 6n +$ ....

$= 6($ ....$)$

marks

 7 Fill in the gaps to prove the following statement.
The difference of the squares of two consecutive integers is always odd. **(3)**

$($ ....$)^2 - ($ ....$)^2 =$ .... $-$ ....

$=$ ....

Hence ....

marks

Make a plan

Had a go 0–4 marks	**Nearly there** 5–9 marks	**Nailed it!** 10–12 marks
Look back at your work on functions and constructing mathematical arguments. Focus on the algebraic skills of substitution and rearrangement in order to tackle the core skills of evaluating functions and finding inverse functions.	Well done! Use the hints in the answers to work out where you could have picked up more marks. Remember that if a question asks you to complete a mathematical argument, you must show every step in your working.	Congratulations! Keep an eye out for more complicated proofs. You might need to work with sums or products of consecutive integers, odd numbers or even numbers.

MY TOTAL MARKS

Answers on page 95

10 Straight-line graphs

  You need to be able to plot straight-line graphs, use the form $y = mx + c$, and find the equation of the line through two points or through one point with a given gradient. You also need to be able to find the equation of a line parallel or perpendicular to a given line.

1 A straight-line graph has equation $y = 3x - 2$
Write down the gradient and y-intercept. ①

Answer: Gradient: ....**3**........ ✓

Answer: y-intercept: ....**−2**........ ✓

$\boxed{1}$
marks

2 A straight line has equation $3x + 2y = 6$
What is the gradient of the line?
Tick **one** box. (1)

 ☑ **A** 3

$y = \dfrac{-3}{2} + 3$

☐ **B** −3

☐ **C** $\dfrac{3}{2}$

☑ **D** $-\dfrac{3}{2}$

$\boxed{0}$
marks

3 The graph shows the relationship between the length of a taxi journey and its cost.

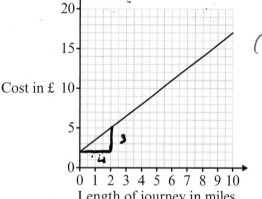

Cost in £

Length of journey in miles

(a) Calculate the gradient of the line. (1)

Answer: .....$\dfrac{3}{4}$........

$\boxed{½}$
marks

(b) Interpret this value. (1)

cost per mile

Answer: ....

$\boxed{0}$
marks

4 The graph shows a straight line, *l*.

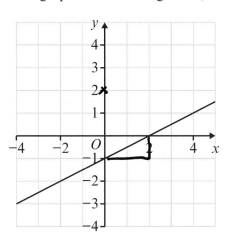

Find the equation of the straight line parallel to *l* that passes through the point (0, 2).
Give your answer in the form $y = mx + c$ (2)

$$y = \frac{1}{2}x - 1$$

Answer: ...$y = \frac{1}{2}x + 2$........................

2 marks

5 A straight line passes through the coordinate points $A(1, 3)$ and $B(3, 8)$.
Find the equation of the line through *A* and *B*.

Core skill Give your answer in the form $y = mx + c$ (2)

$x = 1$ $x = 3$ →   ↑5
$y = 3$ $y = 8$ 2

$m = \frac{5}{2} = 2.5$

Answer: ...$y = 2.5 + \frac{1}{2}$........................

1 marks

6 A straight line, *l*, has equation $y = -\frac{1}{2}x - 1$
Find the equation of the line perpendicular to *l* which passes through point (1, 3).
Give your answer in the form $y = mx + c$ (2)

Answer: $y = 2x + 2 \ +1$........................

1 marks

Make a plan

☑ **Had a go** **0–5 marks**	☑ **Nearly there** **6–8 marks**	☑ **Nailed it!** **9–10 marks**	**6** MY TOTAL MARKS

Make sure you understand what *m* and *c* represent in the equation of a straight line. Then you can tackle the higher core skill of understanding gradients of parallel and perpendicular lines.

Well done! Use the hints in the answers to work out where you could have picked up more marks. Remember that if a question asks you to give your answer in a specific form, you must write it like that.

Congratulations! Keep an eye out for more complicated questions. You might need to solve problems in coordinate geometry using facts about coordinates and straight-line graphs such as finding the equation of a perpendicular bisector.

Answers on page 95

11 Functions and graphs

 You need to be able to plot the graph of a quadratic function and use the graphs of quadratic functions to identify and show the roots, intercepts and turning points of quadratic functions. You also need to be able to use algebra to work out the key points.

1 A graph has equation $y = x^2 + 3x - 2$

(a) Complete the table of values. ①

x	−3	−2	−1	0	1
y	−2	−4	−4	−2	2

✓✓

1 marks

(b) Draw the graph of $y = x^2 + 3x - 2$ ①

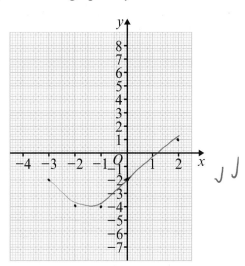

✓✓

1 marks

2 This is the graph of $y = x^2 - 5x + 2$

Core skill

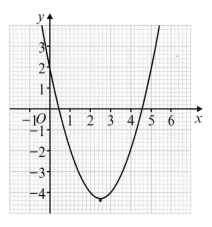

(a) Write down the coordinates of the y-intercept. ①

Answer: (......0.5......, ...2......) ✓

1 marks

(b) Write down the values of x where the graph crosses the x-axis. ①

Answer: $x = ...0.2...$ ✓

Answer: $x = ...4.5...$ ✓

1 marks

(c) Write down the coordinates of the turning point on the graph. ①

Answer: (......2.5......, ...−4.2......) ✓✓

1 marks

3 Jacintha draws the graph of $y = x^2 - 6x + 3$

Complete the working out to find the coordinates of the turning point of Jacintha's graph.

$$x^2 - 6x + 3 = (x \underline{\quad -3 \quad})^2 \underline{\quad -9 \quad} + 3 \checkmark\checkmark$$

$$= (x \underline{\quad -3 \quad})^2 \underline{\quad -6 \quad} \checkmark$$

The coordinates of the turning point are ($\underline{\quad 3 \quad}$, $\underline{\quad -6 \quad}$)

3 marks

4 This is the graph of $y = x^2 - 3x - 2$

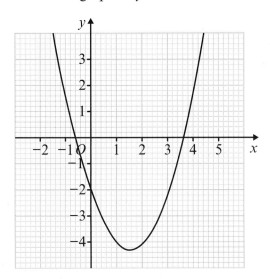

(a) Write down the equation of a suitable straight line that can be used to estimate the solutions to the equation $x^2 - 4x - 1 = 0$

Answer: $y = \underline{\quad x - 1 \quad}$

0 marks

(b) Hence estimate the solutions to the equation $x^2 - 4x - 1 = 0$

Give your answers correct to 1 decimal place.

Answer: $x = \underline{\quad 4.2 \quad}$ or $\underline{\quad -0.2 \quad}$

0 marks

Make a plan

 Had a go
0–5 marks

Make sure you are confident plotting points and reading values from graphs accurately. Then you can tackle the higher core skill of using quadratic graphs and identifying their key features.

 Nearly there
6–8 marks

Well done! Use the hints in the answers to work out where you could have picked up more marks. Remember that if a question asks you to read off a graph, try to be as accurate as possible.

 Nailed it!
9–10 marks

Congratulations! Keep an eye out for more complicated questions about quadratic graphs. You might need to find an approximate solution to an equation using a graph of a function, or find the coordinates of the turning point of a graph given in the form $y = ax^2 + bx + c$

8

MY TOTAL MARKS

Answers on page 96

12 Transformations of graphs

  You need to be able to recognise, sketch and interpret graphs, including quadratic graphs, cubic graphs, reciprocal graphs, exponential graphs and trigonometric graphs. You also need to be able to sketch translations and reflections of a given function.

1 Draw lines to indicate the correct equation for each graph. **(2)**

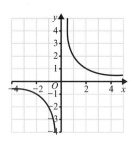 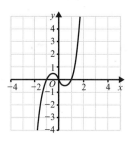 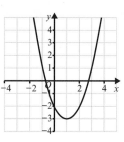 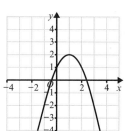

$$y = x^3 - x \qquad y = x^2 - 2x - 2 \qquad y = \frac{2}{x} \qquad y = 1 + 2x - x^2$$

marks

2 The diagram shows the graph of a function $y = f(x)$

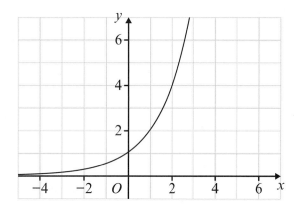

Identify the type of function.
Tick **one** box. **(1)**

☐ **A** Linear
☐ **B** Cubic
☐ **C** Reciprocal
☐ **D** Exponential

marks

3 Sketch the graph of $y = \sin x$ in the interval $0° \leqslant x \leqslant 360°$ **(2)**

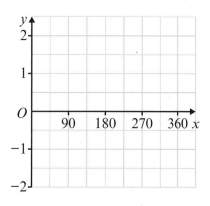

marks

4 The graph of $y = \cos x$ in the interval $-180° \leqslant x \leqslant 180°$ is shown below.

Core skill

On the same set of axes, sketch the graph of $y = -\cos x$ **(1)**

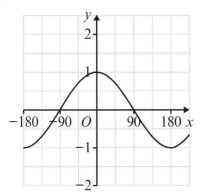

marks

5 The graph of $y = f(x)$ is shown below.

On the same set of axes, sketch the graph of $y = f(-x)$ **(1)**

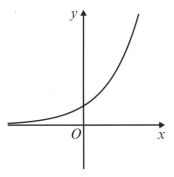

marks

6 Amber draws the graph of $y = x^2$

She then draws two other graphs.

Complete the following sentences: **(4)**

To draw the graph of $y = x^2 + 3$ she should translate the graph ....

by .... units.

To draw the graph of $y = (x - 2)^2$ she should translate the graph ....

by .... units.

marks

Make a plan

Had a go **0–4 marks**	**Nearly there** **5–8 marks**	**Nailed it!** **9–11 marks**	
Make sure you are confident with understanding the shapes of different types of graphs, especially the trigonometry graphs. Then you can tackle the core skill of transforming these graphs.	Well done! Remember to label key points on any sketch graph, and make sure you understand what happens to them under transformations.	Congratulations! Keep an eye out for more complicated questions about transforming graphs. You could be given an unfamiliar function and asked to sketch one or more transformations of the function on the same set of axes.	MY TOTAL MARKS

13 Gradients and areas on graphs

 You need to be able to interpret graphs, including quadratic graphs, cubic graphs, reciprocal graphs, exponential graphs and graphs of non-standard functions. You also need to be able to calculate estimates of gradients and areas under graphs.

1 This is the velocity–time graph for a car journey.

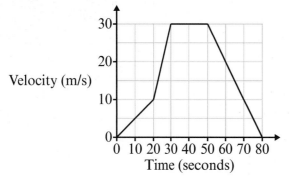

Which of the following statements are true?
Tick **two** boxes. **(2)**

☐ **A** The car accelerated during the first 20 seconds
☐ **B** The car accelerated during the last 30 seconds
☐ **C** The car travelled at a constant speed between 30 and 50 seconds
☐ **D** The car accelerated quickest between 0 and 20 seconds

☐ marks

2 A researcher recorded the number of bacteria in a petri dish at different times after the beginning of an experiment. The graph shows her results.

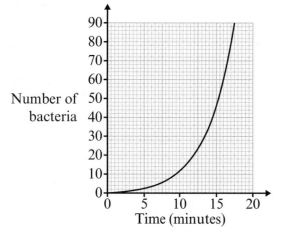

(a) Find the number of bacteria after 15 minutes. **(1)**

Answer: ....

☐ marks

(b) Find, to the nearest minute, the time taken for there to be 60 bacteria. **(1)**

Answer: .... minutes

☐ marks

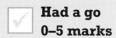

3   Here is the velocity-time graph for a speed walker.

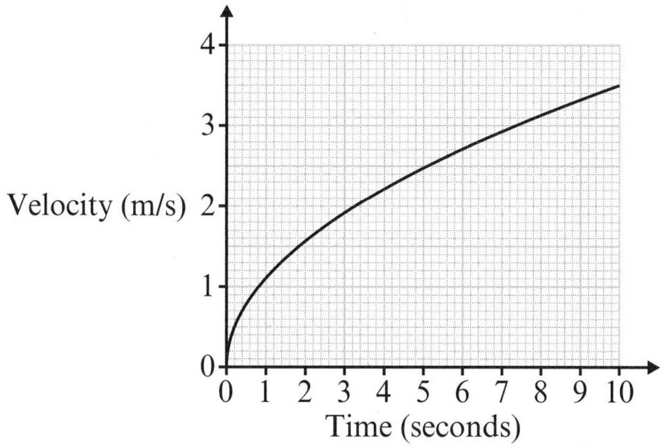

Velocity (m/s)

4

3

2

1

0

0 1 2 3 4 5 6 7 8 9 10
Time (seconds)

(a) Draw a tangent to the curve when $x = 3$ **(1)**   marks

(b) Hence estimate, to 2 decimal places, the acceleration of the walker after 3 seconds.   **(2)**

Answer: .... m/s^2 marks

(c) Using the graph above, and a strip width of 5 units, draw one triangle and one trapezium to represent an estimate of the distance travelled by the walker over these 10 seconds.   **(1)**   marks

(d) Hence calculate an estimate of the distance travelled.   **(2)**

Area under the graph = ... units^2

Answer: .... m   marks

(e) Is your answer to part (b) an overestimate or an underestimate?
Tick **one** box.   **(1)**

☐ **A** Overestimate
☐ **B** Underestimate   marks

My marks

Make a plan

☑ **Had a go**
0–5 marks

Make sure you understand the basic skill of finding a gradient from a straight-line graph. Then you can tackle the core skill of finding an approximation of the gradient of a curve.

☑ **Nearly there**
6–8 marks

Well done! Use the hints in the answers to work out where you could have picked up more marks. Remember that if a question asks you to draw a tangent, you need to use a ruler and draw the line by eye.

☑ **Nailed it!**
9–11 marks

Congratulations! Keep an eye out for more complicated questions about areas under graphs. You could be given a graph representing an unfamiliar context and have to interpret the area under it in context.

MY TOTAL MARKS

Answers
on page 97

14 Equation of a circle

  You need to be able to recognise and use the equation of a circle centred on the origin and find the equation of a tangent to a circle at a given point.

1 A circle has equation $x^2 + y^2 = 25$
Work out the radius of the circle. **(1)**

Answer: ....**5** ✓.......

✓ marks

2 A circle, centred on the origin, has diameter 12 cm.
Work out the equation of the circle.
Tick **one** box. **(1)**

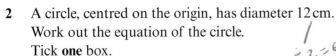
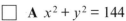

☐ **A** $x^2 + y^2 = 144$
☒ **B** $x^2 + y^2 = 12$
☐ **C** $x^2 + y^2 = 6$
☑ **D** $x^2 + y^2 = 36$

marks

3 A circle, centred on the origin, passes through the point (6, 8).
Calculate the diameter of the circle. **(1)**

Answer = ...**8** ✗...... marks

4 A circle, centred on the origin, has equation $x^2 + y^2 = 20$
Identify the point that lies on the circle.
Tick **one** box. **(1)**

☐ **A** (2, 5)
☒ **B** (5, 2)
☐ **C** (3, 3)
☑ **D** (4, 2)

marks

5 The diagram shows a circle with equation $x^2 + y^2 = 40$, centred on the origin, O.

Core skill

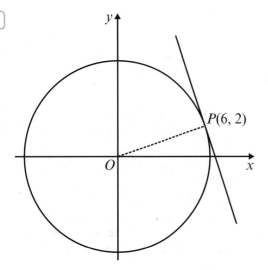

The tangent to the circle at the point P with coordinates (6, 2) is shown.

(a) Calculate the gradient of the radius *OP*. **(1)**

$$\frac{\text{Change in } y}{\text{Change in } x} \qquad \frac{2-0}{6-0} \qquad \frac{2}{6} \ ? \ \checkmark \qquad \frac{1}{3}$$

Answer: m_{OP} = ...

1 marks

(b) Hence calculate the gradient of the tangent to the circle at *P*. **(1)**

Answer: m_{tangent} = ...−3......

0 marks

(c) Hence calculate the equation of the tangent to the circle at *P*.
   Give your answer in the form $y = mx + c$ **(2)**

Answer = ...$y = 3x + 20$..........................

0 marks

 6 A circle has equation $x^2 + y^2 = 17$
   The tangent to the circle at a point *P* has gradient $-\dfrac{1}{4}$

(a) Calculate the gradient of the radius at *P*. **(1)**

Answer: m = ...4......

0 marks

(b) Given that both the *x*- and *y*-coordinates of *P* are positive, find the coordinates of *P*. **(1)**

Answer: (..........1......, ...4..........)

2 marks

Revision Guide pages 19, 20, 31, 32, 33

15 Solving equations

   You need to be able to solve quadratic equations by factorising, using the quadratic formula, or completing the square.

1 If $3x - 2 = x + 8$, what is the value of x?
Tick **one** box. **(1)**

$3x - 2 = x + 8$

$3x - x = 2 + 8$

$2x = 10 \qquad x = \dfrac{10}{2} \qquad x = 5$

- [ ] **A** 3
- [ ] **B** 4
- [x] **C** 5 ✓
- [ ] **D** 6

1 marks

2 What formula gives the solutions to the equation $ax^2 + bx + c = 0$?
Tick **one** box. **(1)**

- [ ] **A** $x = \dfrac{b \pm \sqrt{b^2 - 4ac}}{2a}$
- [ ] **B** $x = \dfrac{b \pm \sqrt{b^2 + 4ac}}{2a}$
- [x] **C** $x = \dfrac{-b \pm \sqrt{b^2 - 4ac}}{2a}$ ✓
- [ ] **D** $x = \dfrac{-b \pm \sqrt{b^2 - 4ac}}{2}$

1 marks

3 Solve the equation $(x + 3)(x - 2) = 0$ **(1)**

$(x + 3) = 0 \rightarrow x = -3$
OR
$(x - 2) = 0 \rightarrow x = 2$

Answer: $x = $ ......−3.... or $x = $ ....2....

1 marks

4 Solve the equation $x^2 + 5x = 0$
Tick **one** box. **(1)**

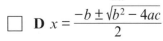 factorise

- [ ] **A** $x = -5$
- [x] **B** $x = 0$ or $x = -5$
- [x] **C** $x = 5$ ✗
- [ ] **D** $x = 0$ or $x = 5$

0 marks

5 Solve the equation $2x^2 + 8x - 5 = 0$
Give your answers correct to 2 decimal places. **(2)**

Core skill

 $x = \dfrac{-8 \pm \sqrt{8^2 - 4 \times 2 \times (-5)}}{2 \times 2}$

$(x^2 + 4x)(x^2 + 4x)$

Answer: $x = $ ...0.55... or $x = $ ...−4.55...

0 marks

6 Complete the working to solve the equation $x^2 = 8x - 15$ **(2)**

Core skill

$$x^2 = 8x - 15$$
$$x^2 - 8x + 15 = 0$$
$$\left(x - \boxed{5}\right)\left(x - \boxed{3}\right) = 0$$ ✓
$$x = \boxed{5} \text{ or } x = \boxed{3}$$ ✓

2 marks

7 $(2x - 1)^2 = x^2 + 5$

(a) Write this equation in the form $ax^2 + bx + c = 0$ **(1)**

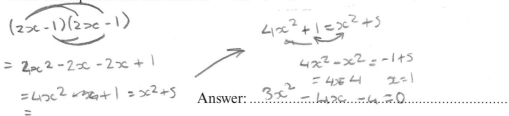

$$(2x - 1)(2x - 1)$$

$$= 2x^2 - 2x - 2x + 1$$

$$= 4x^2 - 4x + 1 = x^2 + 5$$

$$4x^2 + 1 = x^2 + 5$$

$$4x^2 - x^2 = -1 + 5$$

$$= 4x^2 4 \quad x = 1$$

Answer: $3x^2 - 4x - 4 = 0$

 0 marks

(b) Factorise the left-hand side of your answer to part (a) to complete the brackets below. **(1)**

$(3x \ldots +2 \ldots)(x \ldots -2 \ldots) = 0$

0 marks

(c) Solve the equation formed in part (a). **(1)**

Answer: $x = -\dfrac{2}{3}$ or $x = 2$

0 marks

8 (a) Complete the square for this expression: **(1)**

$x^2 + 4x - 3 \equiv \left(x + \boxed{2}\right)^2 - \boxed{7}$

 1 marks

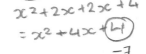

$(x + 2)(x + 2)$

$x^2 + 2x + 2x + 4$

$= x^2 + 4x + \boxed{4}$

-7

(b) Solve the equation $x^2 + 4x - 3 = 0$
Tick **one** box. **(1)**

☐ **A** $x = 2 \pm \sqrt{7}$
☐ **B** $x = 4 \pm \sqrt{3}$
☑ **C** $x = -2 \pm \sqrt{7}$ ✓
☐ **D** $x = 2 \pm \sqrt{5}$

$(x + 2)^2 - 7 = 0$

$(x + 2)^2 = 7$

$x + 2 = \pm\sqrt{7}$

$x = -2 \pm \sqrt{7}$

1 marks

Make a plan

✓ **Had a go** **0–5 marks**	✓ **Nearly there** **6–10 marks**	✓ **Nailed it!** **11–13 marks**
Make sure you are confident with the basic skill of factorising quadratic expressions. Then you can tackle the core skill of solving quadratic equations in different contexts.	Well done! Remember that if a question asks you to solve a quadratic equation to a given degree of accuracy then you should use the quadratic formula.	Congratulations! Keep an eye out for quadratic equations in geometry and probability questions as well – you might need to pick which of the two solutions is appropriate for a particular question.

7
MY TOTAL MARKS

Answers on page 99

16 Simultaneous equations and iteration

  You need to be able to solve two linear simultaneous equations and simultaneous equations where one of the equations involves a squared term. You also need to be able to find approximate solutions to equations using iteration.

1 $x + 2y = 5$
$2x + 3y = 8$
What pair of numbers solve these simultaneous equations?
Tick **one** box. **(1)**

☐ **A** $x = 5, y = 0$
☐ **B** $x = -1, y = 3$
☐ **C** $x = 3, y = 1$
☐ **D** $x = 1, y = 2$

marks

2 Solve these simultaneous equations **(2)**

Core skill

$2x - 3y = 5$
$3x + 2y = 14$

Answer: $x = $ .... , $y = $ ....

marks

3 $2x + y = 9$
$x^2 - x - 21 = y$
What pair of numbers solve these simultaneous equations?
Tick **one** box. **(1)**

☐ **A** $x = 2, y = 5$
☐ **B** $x = -1, y = 11$
☐ **C** $x = 3, y = 3$
☐ **D** $x = 5, y = -1$

marks

4 Solve these simultaneous equations. **(3)**

Core skill

$y = x + 4$
$x^2 + y^2 = 26$
Fill in the gaps to help you.

$$x^2 + (....................)^2 = 26$$

$$x^2 + ... = 26$$

$$x^2 + ... = 0$$

Answer: $x = $ .... , y ....

or $x = $ .... , y ....

marks

5 The equation $x^3 - 2x + 1 = 0$ can be rearranged to give $x = \dfrac{x^3 + a}{b}$ where a and b are numbers.

Core skill

(a) Find the values of a and b. **(1)**

Answer: $a =$ .... , $b =$ ....

marks

(b) Using $x_{n+1} = \dfrac{x_n^3 + a}{b}$ with $x_0 = 0$ find the values of x_1, x_2 and x_3 **(1)**

Answer: $x_1 =$ ....

Answer: $x_2 =$ ....

Answer: $x_3 =$ ....

marks

(c) Use your answer to part (b) to predict a solution to the original equation correct to 1 decimal place. **(1)**

Answer: $x =$ ....

marks

Make a plan

☑ **Had a go**
  0–4 marks

Make sure you are confident with the basic skills of rearranging equations and substituting into expressions. Then you can tackle the core skill of solving simultaneous equations by substitution.

☑ **Nearly there**
  5–7 marks

Well done! Use the hints in the answers to work out where you could have picked up more marks. You can check answers to simultaneous equations by substituting them back into the original equations.

☑ **Nailed it!**
  8–10 marks

Congratulations! Remember that in non-linear simultaneous equations you usually need to find two sets of x- and y-values and pair them up correctly.

MY TOTAL MARKS

Answers on page 100

17 Using equations and inequalities

You need to be able to translate situations into algebraic expressions, formulae or equations and solve these equations. You also need to be able to solve linear inequalities and quadratic inequalities and represent their solutions on a number line or graph.

1 Solve the inequality $4x - 7 \geqslant 21$
Tick **one** box. **(1)**

- [ ] **A** $x \leqslant 7$
- [ ] **B** $x \geqslant 7$
- [ ] **C** $x \geqslant 3.5$
- [ ] **D** $x \leqslant 3.5$

2 2 adult cinema tickets plus 4 child cinema tickets cost £31
4 adult cinema tickets plus 5 child cinema tickets cost £50
Let x represent the cost of an adult ticket and y the cost of a child ticket.

(a) Write down two equations connecting x and y. **(1)**

Answer: ....

and ....

(b) Which of the following are the prices of the adult and child tickets?
Tick **one** box. **(1)**

- [ ] **A** Adult: £5.50; Child: £5
- [ ] **B** Adult: £6.50; Child: £4.50
- [ ] **C** Adult: £7.50; Child: £4
- [ ] **D** Adult: £8.50; Child: £3.50

3 On the graph, draw the region of points whose coordinates satisfy these inequalities. **(2)**
$x \geqslant -1$ \qquad $y > 2$ \qquad $x + y < 7$

My marks ⌄

□ marks

□ marks

□ marks

□ marks

4 Calculate the size of the largest angle in this triangle. (2)

Triangle with angles labelled $2x$ (top), $x + 10$ (bottom left), and $3x - 40$ (bottom right).

Answer: ....°

5 (a) Factorise $x^2 - 6x + 8$ (1)

Core skill

$x^2 - 6x + 8 = ($ .... $)($ .... $)$

(b) Hence, or otherwise, solve the inequality $x^2 - 6x + 8 < 0$ (1)

Answer: ....

(c) Show your answer on a number line. (1)

Number line from −5 to 5.

6 Given that the area of this rectangle is $78\,\text{cm}^2$, find the value of x correct to 2 decimal places. (2)

Rectangle with sides labelled $2x + 1$ cm (right side) and $4x - 3$ cm (bottom).

Answer: $x =$ ....

My marks

marks

marks

marks

marks

marks

Make a plan

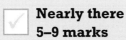 **Had a go**
0–4 marks
Practise solving linear inequalities and representing the solutions on number lines. Make sure you check that your final answer satisfies the original inequality.

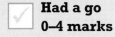 **Nearly there**
5–9 marks
Well done! Use the hints in the answers to work out where you could have picked up more marks. If you need to solve a quadratic inequality, it can help to sketch a graph to see which parts of the curve are above or below the x-axis.

Nailed it!
10–12 marks
Congratulations! You will be able to make use of your algebra skills if you have to form and solve inequalities in probability, geometry or ratio and proportion questions.

 MY TOTAL MARKS

Answers on page 100

18 Sequences

You need to be able to continue a given sequence, find missing terms in a sequence, recognise different types of sequence and solve problems involving sequences.

1   The rule for generating a sequence is 'add two consecutive terms to get the next term'.
The first four terms are
4, 5, 9, 14
Write down the next two terms of this sequence.   **(2)**

Answer: .... and ....   marks

2   The nth term of a sequence is $25 - n^2$
Which one of these numbers is **not** a term in the sequence?
Tick **one** box.   **(1)**

☐ **A** 16
☐ **B** 9
☐ **C** 25
☐ **D** 0
                                                                                   marks

3   The nth term of a sequence is $5n + 3$

(a) Work out the ninth term of the sequence.   **(1)**

Answer: ....   marks

(b) Work out the first term of the sequence that is greater than 61   **(1)**

Answer: ....   marks

4   The nth term of an arithmetic sequence is given by $u_n = 4n - 3$

Core skill   (a) Is 78 a term in this sequence?   **(1)**

☐ **A** Yes
☐ **B** No
                                                                                   marks

(b) Write an expression for u_n in terms of u_{n-1} **(1)**

Answer: $u_n =$ ....   marks

5 The nth term of a sequence is given by $u_n = 2^n$

Core skill

(a) Fill in the blank in the following sentence. (1)

The name given to a sequence of this type is a .... sequence.

(b) Write down the first three terms of the sequence. (1)

Answer: ...., ... and ....

6 The first five terms of a sequence are
1, 7, 25, 79, 241
Work out an expression for the nth term.
Tick **one** box. (1)

☐ **A** $3n - 2$
☐ **B** $n^3 - 2$
☐ **C** 3^{n-2}
☐ **D** $3^n - 2$

7 The first four terms of a geometric sequence are
$\sqrt{2}$, k, $8\sqrt{2}$, 32
Work out the value of k.
Tick **one** box. (1)

☐ **A** 4
☐ **B** $2\sqrt{2}$
☐ **C** 6
☐ **D** $4\sqrt{2}$

My marks

☐ marks

☐ marks

☐ marks

☐ marks

Make a plan

☑ **Had a go**
0–4 marks
Try revising sequence problems, and focus on the core skills of recognising different types of sequences and generating terms.

☑ **Nearly there**
5–8 marks
Well done! Use the hints in the answers to work out where you could have picked up more marks. Remember that if a question asks you to determine if a given number is a term in a sequence, you can try a few values for n.

☑ **Nailed it!**
9–11 marks
Congratulations! Keep an eye out for sequences with unknown terms – you might need to find more than one missing value using a general rule.

☐ **MY TOTAL MARKS**

Answers on page 101

19 Finding *n*th terms

You need to be able to find an expression for the *n*th term for both linear and quadratic sequences.

1 A sequence has *n*th term $5n - n^2 + 1$. What type of sequence is this?
Tick **one** box. **(1)**

☐ **A** A linear sequence
☐ **B** A quadratic sequence
☐ **C** A geometric sequence
☐ **D** A Fibonacci sequence

2 The first five terms of a linear sequence are shown below. Fill in the blanks to work out the *n*th term of the sequence. **(2)**

Zero term

☐ 15 23 31 39 47

+ ☐ + ☐ + ☐ + ☐

*n*th term = ☐ *n* + ☐

3 The first five terms of a sequence are
5, 8, 11, 14, 17
Write down an expression for the *n*th term. **(2)**

*n*th term = ... *n* + ...

Core skill

4 Here are the first four terms of a sequence.
10, 7, 4, 1
Work out an expression for the *n*th term.
Tick **one** box. **(1)**

Core skill

☐ **A** $3n + 7$
☐ **B** $n - 3$
☐ **C** $13 - 3n$
☐ **D** $10 - 3n$

5 The first four terms of a sequence are
0, 3, 8, 15
Work out the *n*th term of this sequence.
Tick **one** box. **(1)**

☐ **A** n^2
☐ **B** $3n - 3$
☐ **C** $n + 3$
☐ **D** $n^2 - 1$

My marks

☐ marks

☐ marks

☐ marks

☐ marks

☐ marks

6 The first five terms of a sequence are

6, 11, 18, 27, 38

Write down an expression for the nth term. **(2)**

nth term = $n^2 +$ .... $n +$ ....

marks

7 The first five terms of a sequence are

3, 9, 19, 33, 51

Write down an expression for the nth term. **(2)**

Answer: ....

marks

8 The first five terms of a sequence are

$-1.5, -5, -7.5, -9, -9.5$

Write down an expression for the nth term. **(3)**

nth term = ... n^2 .... n ....

marks

Make a plan

✓ **Had a go**
0–4 marks

Recap your work on sequences. Focus on the core skill of finding an expression for the nth term of a linear sequence.

✓ **Nearly there**
5–8 marks

Well done! Remember that if a question asks you to find an expression for the nth term of a quadratic sequence, you should look at the second difference.

✓ **Nailed it!**
9–14 marks

Congratulations! Keep an eye out for quadratic sequences made from patterns – you might need to find an expression for the nth term of a quadratic sequence in practical situations.

MY TOTAL MARKS

Answers on page 102

Revision Guide pages 65, 67, 91, 92

20 Units, scale drawings and maps

10 🖩 You need to be able to work confidently with standard units and related compound units in numerical contexts. You also need to be able to use scale factors, scale diagrams and scales and bearings on maps.

My marks ⌄

1 The average speed of a cyclist is 10 m/s.
Convert 10 m/s to km/h.
You must show your working. **(1)**

....

10 m/s = ... m/h

Answer: .... km/h ☐ marks

2 A car is 4.8 m long. A scale model of the car is made using a scale of 1 : 32
What is the length of the scale model?
Give your answer in centimetres.
You must show your working. **(2)**

Answer: .... cm ☐ marks

3 The bearing of Coventry from Stratford-upon-Avon is 031°
Find the bearing of Stratford-upon-Avon from Coventry.
Tick **one** box. **(1)**

☐ **A** 121°
☐ **B** 329°
☐ **C** 211°
☐ **D** 031°

☐ marks

4 The scale diagram shows the position of two towns A and B.

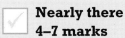

The scale of the diagram is $1 : 500\,000$
A shopping centre S, is on a bearing of 060° from A and 310° from B

(a) On the diagram mark the position of the shopping centre S with a cross (x).
Label it S. **(2)** marks

(b) Use the scale to work out the distance between town A and the shopping centre S.
Give your answer in km correct to 2 significant figures. **(2)**

Answers: .... km marks

5 The average fuel consumption of a car is 42 miles per gallon.
What is the fuel consumption in km per litre?
Use 5 miles = 8 km
1 gallon = 4.55 litres
Give your answer correct to 3 significant figures.
You must show your working. **(3)**

42 miles = 42 ÷ ☐ × ☐ = ☐ km

Answers: .... km/litre marks

21 Ratio

  You need to be able to express one quantity as a fraction of another, where the fraction is less than 1 or greater than 1. You also need to be able to use ratio notation, including reduction to simplest form.

1 A ball pool in a play centre has a mix of red balls and white balls.
$\frac{5}{9}$ of the balls are red.

(a) What is the ratio of the number of red balls to the number of white balls in its simplest form? **(1)**

$\frac{5}{9}$ are red so $\boxed{\frac{4}{9}}$ are white

ratio of red : white is $\boxed{5}$: $\boxed{4}$

 1 marks

(b) The total number of balls in the ball pool is less than 1000.
What is the greatest number of red balls and the greatest number of white balls that are in the ball pool? **(2)**
Largest possible number of balls = ...999..........

$$5+4=9 \qquad \frac{1000}{9}$$

Answer: red: ....

Answer: white: ....

 marks

2 Emma and Matt win some money in a raffle.
They share the money in the ratio 3 : 5
Matt gets £25 more than Emma.
How much money did they win in total?
You must show your working. **(3)**

 Core skill

£25 = $\boxed{3}$ parts

£ $\boxed{12.50}$ = 1 part

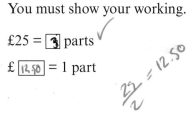

Answer: £ ....

 2 marks

3 A box contains 56 cards.

Each card is either white or black.

The ratio of the number of white cards to the number of black cards is <u>1 : 1</u>

7 white cards are removed from the box.

Find the ratio of the number of white cards to the number of black cards now in the box.

Give your answer in its simplest form. (3)

At the start:

number of white cards = [98 28] ✓

number of black cards = [28] ✓

After 7 white cards removed: 21 ✓

21 : 28

Answer: ....21.......

 2 marks

4 A large box of chocolates has a mix of milk chocolates, dark chocolates and white chocolates.

The ratio of the number of milk chocolates to the number of dark chocolates is 3 : 4

The ratio of the number of dark chocolates to the number of white chocolates is 5 : 1

There are fewer than 80 chocolates in the box.

What is the greatest possible number of white chocolates in the box? (3)

milk	:	dark	:	white	
3	:	4			×5 ✓
		5	:	1	× 4 ✓
15	:	20	:	4	

15 + 20 + 4   80/39   ✓
39

8

Answer: ...15 : 20 : 4...

 3 marks

Make a plan

☑ **Had a go**
0–4 marks

Ratios can be tricky. Recap your basic ratio skills like finding equivalent ratios and dividing an amount in a given ratio. You might also need to revise calculations with fractions.

☑ **Nearly there**
5–8 marks

Well done! Make sure you know how to convert ratios into fractions, and read questions carefully to make sure your answers make sense.

☑ **Nailed it!**
9–12 marks

Congratulations! Understanding ratios can help you solve problems involving similar shapes, maps and scale drawings.

8
MY TOTAL MARKS

Answers
on page 103

Revision Guide page 61

22 Ratio (continued)

10

You need to be able to understand and use proportion in terms of ratios. You also need to be able to relate ratios to fractions and solve ratio problems involving algebra.

My marks

1 Becca makes pastry by mixing flour and butter in the ratio 3 : 2

Core skill

(a) She has 250 grams of butter and plenty of flour. What is the maximum weight of pastry she can make?
You must show your working. **(1)**

250 g = 2 parts
125 g = 1 part

250 = 2

$\frac{250}{2}$ = 125

125 × 3

375

Answer: 375 g

marks

(b) Write an equation to show the linear relationship between the amount of flour, F grams in terms of the amount of butter, B grams.
You must show your working. **(2)**

$\frac{F}{B} = \frac{3}{2}$ so F = 1.5

Answer: 1.5

marks

2 Anna follows these instructions to make a drink.
Add 100 ml of cordial to 0.5 litres of water.
She uses 3 litres of water to make the drink.
How much drink has she made? **(3)**

0.5 litres = 500 ml
cordial : water = 1 : 5

3 ÷ 5 = 0.6

Answer: 0.6 l

marks

3 y and x are related by the equation $4y = 5x$
Write down the ratio of $y : x$ (2)

Answer: .... 4:5 ....

 2 marks

 4 The ratio $2x : x - 2$ is equivalent to the ratio $10x : 2x - 1$
Find the possible values of x.
You must show your working. (4)

$$\frac{2x}{x-2} = \boxed{\frac{10x}{2x-1}}$$

$$2x(2x-1) = 10x(x-2)$$
$$4x^2 + 2x - 1 = 10x^2 + 10x - 2$$
$$4x^2 - 2x = 10x^2 - 20x$$

Answer: ....

marks

Make a plan

☑ **Had a go**
0–4 marks
Practise simple proportion questions, and make sure you are confident rearranging equations involving fractions.

☑ **Nearly there**
5–8 marks
Well done! Read questions carefully, and don't be afraid to turn word problems into algebra problems.

☑ **Nailed it!**
9–12 marks
Congratulations! Remember that ratio problems can sometimes lead to tricky algebra like solving quadratic equations – be confident in your skills and check that your final answers make sense.

MY TOTAL MARKS

Answers on page 104

23 Percentages

⏱ **10** 🖩 You need to be able to define percentage as 'number of parts per hundred'. You also need to be able to express one quantity as a percentage of another and work out percentage changes.

My marks ⌄

1 Express 36 as a percentage of 48
You must show your working. **(1)**

$$\frac{\boxed{}}{\boxed{}} \times \boxed{}$$

Answer: .... % ☐ marks

2 Work out the multiplier for an increase of 8.5% followed by a decrease of 12% **(2)**

multiplier for increase of 8.5% = ☐
multiplier for decrease of 12% = ☐

Answer: .... ☐ marks

3 Wayne sells radios.
In one week he sells 96 radios.
The next week he sells 132 radios.
Core skill Work out the percentage increase in the number of radios he sells.
You must show your working. **(3)**

Actual increase = ☐ – ☐

Answer: .... ☐ marks

My marks

4 Gavin invests £600 for 4 years in a bank account.
The account pays simple interest at a rate of 2.4% per year.
Work out the total amount of interest Gavin has received at the end of 4 years. **(3)**

Interest for 1 year = $\dfrac{\boxed{}}{\boxed{}} \times \boxed{}$

Interest for 4 years =

Answer: £ ....

marks

5 In 2018 Sarah was given an increase of 2.5% on her salary for 2017.
In 2019 Sarah was given an increase of 3.2% on her salary for 2018.
In 2019 her salary was £26 445
What was Sarah's salary in 2017? **(3)**

Answer: £ ....

marks

24 Compound measures

 10 You need to be able to use compound units such as speed, rates of pay, unit pricing, density and pressure.

1 The water from a garden hose flows at a rate of 75 litres per minute.
A children's paddling pool holds 530 litres of water.
How long will it take to fill the paddling pool in minutes and seconds?
You must show your working. **(2)**

Time = [ ] ÷ [ ] minutes

Answer: .... minutes .... seconds

marks

2 Cora travels 63 km in 2 hours 15 minutes.
Work out her average speed in km/h
You must show your working. **(3)**

Core skill

2 hours 15 minutes = [ ] hours

Answer: .... km/h

marks

3 A box exerts a force of 120 newtons on a table.
The pressure on the table is 25 newtons/m².
Calculate the area of the box that is in contact with the table.
You must show your working.

$$p = \frac{F}{A}$$
p = pressure
F = force
A = area

(3)

Answer: ....

marks

4 The density of orange juice is 1.25 grams per cm³.

The density of lemon juice is 1.03 grams per cm³.

The density of sparkling water is 0.99 grams per cm³.

220 cm³ of orange juice are mixed with 180 cm³ of lemon juice and 750 cm³ of sparkling water to make a drink with a volume of 1150 cm³.

Work out the density of the drink.

Give your answer correct to 2 decimal places.

You must show your working.

(4)

Mass of orange juice = ☐ × ☐ = ...

Mass of lemon juice = ☐ × ☐ = ...

Mass of sparkling water = ☐ × ☐ = ...

Total mass = ...

Answer: ....

Make a plan

Had a go
0–4 marks

Remember that if a quantity is expressed as something **per** something, the quantity is a compound measure. Revise the formula triangles for compound measures.

Nearly there
5–8 marks

Well done! Check that you are using consistent units whenever you are calculating with compound measures.

Nailed it!
9–12 marks

Congratulations! Watch out if you have to convert a compound measure. You will usually need to do more than one calculation.

MY TOTAL MARKS

Answers on page 106

Revision Guide pages 68, 69, 70

25 Proportion

  You need to be able to solve problems involving direct and inverse proportion, including graphical and algebraic representations. You also need to be able to recognise and interpret graphs that illustrate direct and inverse proportion. In addition, you should be able to construct and interpret equations that describe direct and inverse proportion.

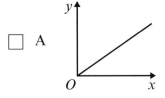

1 Which equation describes a direct proportion relationship between y and x? **(1)**
Tick **one** box.

- [ ] **A** $y = x + 3$
- [ ] **B** $y = x^2$
- [ ] **C** $y = \dfrac{5}{x}$
- [ ] **D** $y = 2x$

My marks ⌄

[ ] marks

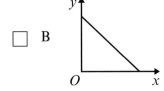

2 Which of the following graphs shows the relationship 'y is inversely proportional to x'? **(1)**
Tick **one** box.

- [ ] A

- [ ] B

- [ ] C

- [ ] D

[ ] marks

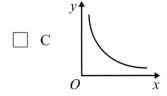

3 a varies inversely as the square of b
If the value of b is doubled, what happens to the value of a?
Tick **one** box. **(1)**

- [ ] **A** It is multiplied by 2
- [ ] **B** It is divided by 2
- [ ] **C** It is multiplied by 4
- [ ] **D** It is divided by 4

 [ ] marks

My marks ⌄

4 Paul exchanged 200 British pounds for 232 euros.
Work out the exchange rate from British pounds to euros. **(2)**

1 British pound = ... euros

Answer: ....

☐ marks

5 It takes 8 men 9 days to complete a job.
Work out how long it would take 3 men to complete the same job. **(3)**

1 man would take ☐ × ☐ days

Answer: .... days

☐ marks

6 x is inversely proportional to the square of y
When $y = 6$, $x = 7.5$

Core skill

(a) Find a formula for x in terms of y
You must show your working. **(3)**

$x \propto \dfrac{\Box}{\Box}$

so $x = \dfrac{\Box}{\Box}$

Answer: ....

☐ marks

(b) Calculate the value of x when $y = 5$
You must show your working. **(1)**

When $y = 5$, $x = \dfrac{\Box}{\Box}$

Answer: ....

☐ marks

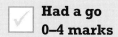
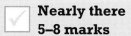
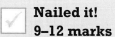

26 Rates of change

  You need to be able to interpret the gradient at a point on a curve as the instantaneous rate of change. You also need to apply the concepts of rate of change in numerical, algebraic and graphical contexts.

 1 The graph shows the depth of water, d cm in a tank after t seconds.

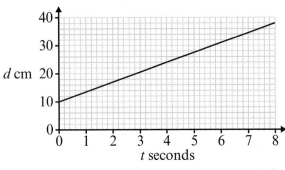

(a) Work out the gradient of the straight line. **(2)**

gradient = $\dfrac{\boxed{}}{\boxed{}}$

Answer: .... marks

(b) What does the gradient of the line represent? **(1)**

Answer: .... marks

 2 The diagram shows part of the graph of $y = 3x^2 - 3x + 5$

Core skill

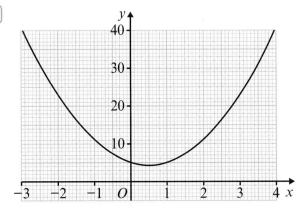

P is the point on the graph where $x = -2$
Calculate an estimate of the gradient of the graph at the point P. **(3)**

Answer: m_p = ... marks

3   The graph gives information about the velocity of a particle from 0 to 40 seconds.

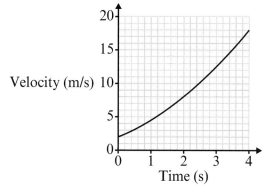

(a) What is the gradient of the graph after 2 seconds?   **(2)**

Answer: ....

 marks

(b) What does the gradient of the graph represent?   **(1)**

Answer: ....

marks

4   The graph shows the temperature of a cup of coffee from 0 to 150 seconds.

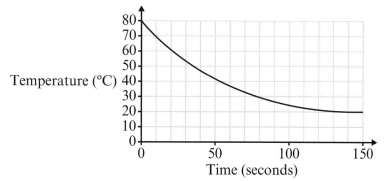

Work out the average rate of decrease of temperature between $t = 0$ and $t = 150$ **(2)**

Answer: ....

 marks

Make a plan

☑ **Had a go**
0–4 marks

Try practising more questions on gradients of straight lines.
Make sure you are confident about finding the right parts of graphs to work out gradients.

☑ **Nearly there**
5–8 marks

Well done! Remember to read the scales carefully when you are using triangles to work out gradients on graphs.

☑ **Nailed it!**
9–11 marks

Congratulations! Keep an eye out for interpreting graphs when set in real-life contexts. Make sure your final answers make sense in the context of the question.

MY TOTAL MARKS

Answers on page 107

27 Growth and decay

  You need to be able to set up, solve and interpret the answers in growth and decay problems, including compound interest. You also need to be able to work with general iterative processes.

1 A savings account pays 3% per annum compound interest.
Nisha puts £500 into the account and does not make any withdrawals.
Complete the working to find the total amount in the account after 4 years. **(2)**

Core skill

Amount after 4 years = 500 × ☐^☐

Answer: £ .... marks

2 The value of a car £V is given by
$V = 30\,000 \times 0.9^n$
where n is the age of the car in complete years.

(a) What is the value of the car after 3 complete years? **(1)**

$V = 30\,000 \times 0.9^{☐}$

Answer: £ .... marks

(b) After how many complete years will the value of the car be less than £16 000? **(2)**

Answer: .... years marks

3 The population of a town is modelled by the equation $P = 4200 \times 1.04^n$, where n is the number of years after 2010.

(a) What was the population of the town in 2010? **(1)**

Answer: ....

(b) Use the model to find an estimate for the population, P, of the town in 2025.
You must show your working. **(2)**

$n = ☐$

Answer: $P = $ .... marks

4 **(a)** Show that the equation $x^3 + 5x = 2$ has a solution between $x = 0$ and $x = 1$ **(2)**

Core skill

Let f(x) = ...

f(0) = ...

f(1) = ...

....

....

marks

The equation $x^3 + 5x = 2$ can be rearranged to give $x = \dfrac{2}{5} - \dfrac{x^3}{5}$

(b) Starting with $x_0 = 0$, use the iteration formula $x_{n+1} = \dfrac{2}{5} - \dfrac{x_n^3}{5}$ twice to find an estimate for the solution of $x^3 + 5x = 2$ **(2)**

Answer: ....

marks

Make a plan

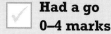 **Had a go**
0–4 marks

Always check the sense of your answers with growth and decay questions – should they be bigger or smaller than values given in the question? Make sure you know how to use your calculator so you can do iteration questions quickly and accurately.

 Nearly there
5–8 marks

Well done! Use the hints in the answers to work out where you could have picked up more marks. You need to be confident about working with percentages, multipliers and indices so have a look at all these topics.

 Nailed it!
9–12 marks

Congratulations! Keep an eye out for working with growth and decay in real-life contexts that may require interpretation.

 MY TOTAL MARKS

 Answers on page 109

28 Angle problems

 You need to be able to apply the properties of angles, including angles in polygons, and be able to solve problems involving angles.

 1 **(a)** Work out the size of the angle x in the diagram.
Give your reasons in full. **(2)**

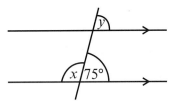

Answer: $x =$ ....$°$ marks

Reason: ....

(b) Work out the size of the angle y in the diagram. **(2)**

Answer: $y =$ ....$°$ marks

Reason: ....

 2 A regular octagon is shown below.

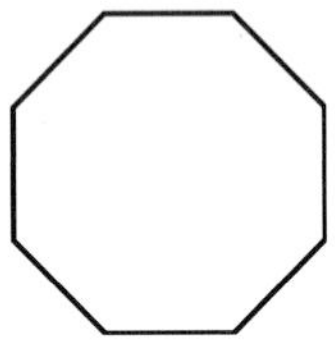

(a) Write down the size of one exterior angle. **(1)**

Answer: ....$°$ marks

(b) Calculate the size of one interior angle. **(1)**

Answer: ....$°$ marks

My marks ⌄

3 The sum of the interior angles of a polygon is 2700°.
Calculate the number of sides. **(1)**

Answer: .... marks

4 A triangle has three angles equal to $(2x - 10)°$, $(3x + 20)°$, and $(4x - 30)°$
Calculate the exact size of the largest angle in the triangle. **(2)**

Answer: ....° marks

5 **(a)** Calculate the size of angle x in the diagram below. **(1)**

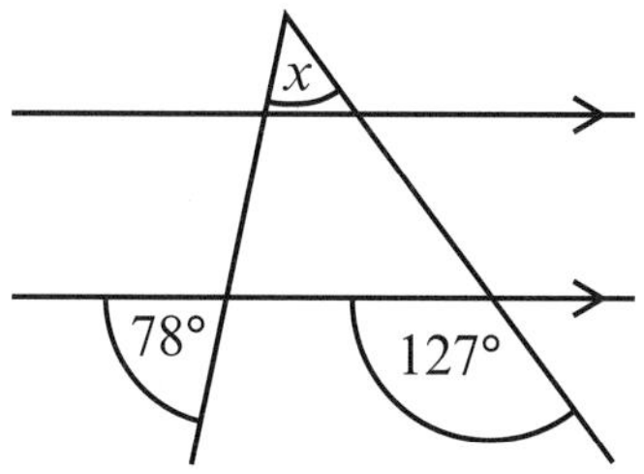

Answer: $x = $ ....° marks

(b) Explain your reasoning. **(1)**

....

....

.... marks

Make a plan

Had a go 0–4 marks	**Nearly there** 5–8 marks	**Nailed it!** 9–11 marks
Make sure that you understand the correct terminology for angle rules. Then you can tackle the core skill of giving reasons for your answers.	Well done! Use the hints in the answers to work out where you could have picked up more marks. Remember that if a question asks you to give a reason for your answer, you must use the correct terminology.	Congratulations! Keep an eye out for multistep questions involving angle properties and questions that ask you to prove the size of a given angle, or to work algebraically.

MY TOTAL MARKS

Answers on page 110

29 Constructions

  You need to be able to use standard ruler and compass constructions to solve problems in loci and draw plans and elevations of 3-D shapes.

1 Construct the locus of points that are within 5 cm of point *A* and 4 cm of point *B*. **(2)**

My marks

A
●

B
●

marks

2 *ABCD* represents a rectangular garden. *AB* = 10 m and *BC* = 18 m.
1 cm in the diagram represents 2 m.

Core skill

Tim wants to plant a crop of beans within 8 m of *C* and within 6 m of side *AD*.
Construct the region in which Tim should plant his broad beans. **(3)**

marks

3 Use the grid to draw the side elevation, front elevation and plan of this object. (3)

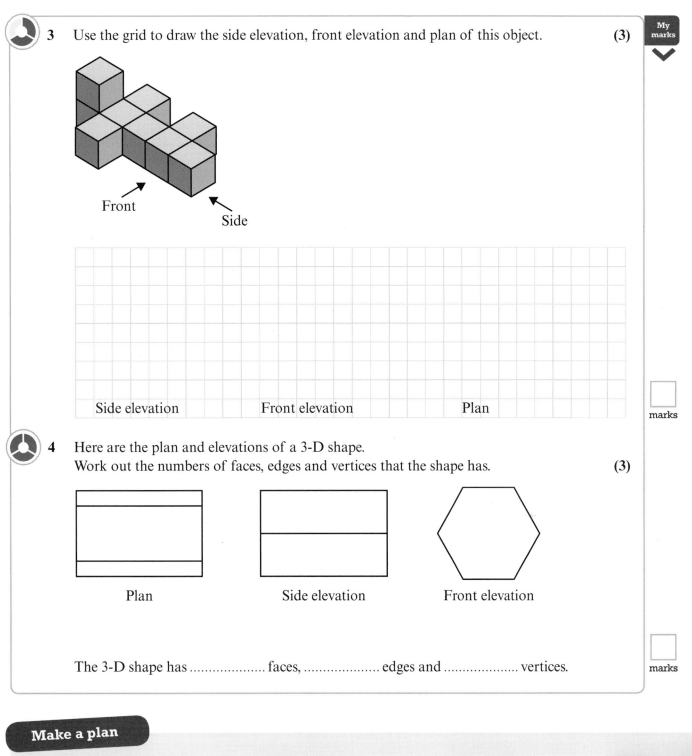

Front

Side

Side elevation   Front elevation   Plan

marks

4 Here are the plan and elevations of a 3-D shape.
Work out the numbers of faces, edges and vertices that the shape has. (3)

Plan   Side elevation   Front elevation

The 3-D shape has .... faces, ... edges and .... vertices.

marks

Make a plan

☑ **Had a go**
0–4 marks

Make sure that you are confident with standard ruler and pair of compasses constructions.
Then you can tackle the core skill of constructing loci.

☑ **Nearly there**
5–8 marks

Well done! Use the hints in the answers to work out where you could have picked up more marks. Remember that if a question asks you to construct a locus, you must show all of your construction lines.

☑ **Nailed it!**
9–11 marks

Congratulations! Keep an eye out for problems where you have to construct more than two loci and where you have to recognise shapes from their plans and elevations.

MY TOTAL MARKS

Answers
on page 110

30 Congruence and similarity

  You need to be able to use similarity and the rules for congruence to find sides and angles, and construct proofs of congruency.

1 Which of the following is **not** a correct criterion for congruence?
Tick **one** box. **(1)**

☐ **A** RHS
☐ **B** SSS
☐ **C** SAS
☐ **D** SSA

☐ marks

2

Core skill

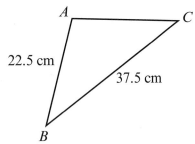 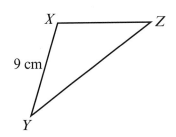

$AB = 22.5\,\text{cm}$
$BC = 37.5\,\text{cm}$
$XY = 9\,\text{cm}$

Triangle XYZ is similar to triangle ABC.

(a) Work out the ratio of the length of AB to the length of XY in its simplest form. **(1)**

$AB:XY = \boxed{} : \boxed{}$

Answer: ....

☐ marks

(b) Work out the length of YZ. **(1)**

$YZ = \dfrac{\boxed{}}{\boxed{}}$

Answer: .... cm

☐ marks

3 Triangles ABC and DEF are similar. Side DE corresponds to side AB and side EF corresponds to side BC.

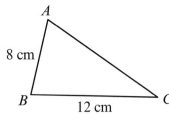 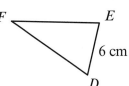

$AB = 8\,\text{cm}$, $DE = 6\,\text{cm}$ and $BC = 12\,\text{cm}$.

(a) Write down the angle that corresponds to angle A. **(1)**

Answer: ....

☐ marks

(b) Calculate the length EF. **(1)**

Answer: .... cm

☐ marks

4   In the diagram below, $BC = CD$.

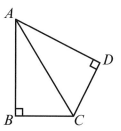

State four facts to prove that triangles ABC and ADC are congruent.   **(4)**

1: ....

2: ....

3: ....

4: ....

marks

5   Triangles ABC and DEF are similar. Side DE corresponds to side AB and side EF corresponds to side BC.

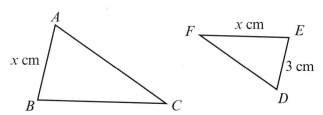

Write down, in term of x, the length of side BC.   **(2)**

Ratio of corresponding sides: ....

Answer: $BC =$ .... cm   marks

31 Transformations

 You need to be able to identify, describe and construct transformations, including describing invariance.

1

Core skill

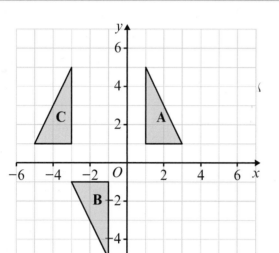

Describe the single transformation that maps:

(a) triangle **A** onto triangle **B**   **(2)**

Rotation (180°)

Answer: ..~~Reflection~~.. over ....(0,0)............

2 marks

(b) triangle **A** onto triangle **C**.   **(2)**

Answer: ...Reflection.. over.. x=-1........ ✓

2 marks

2

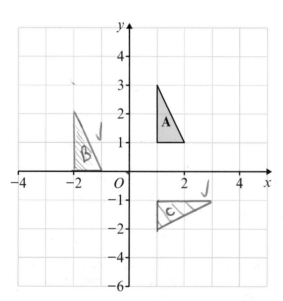

Triangle **B** is a translation of triangle **A** with vector $\begin{pmatrix} -3 \\ -1 \end{pmatrix}$.

Triangle **C** is a rotation of triangle **A** through 90° clockwise about (0, 0).
Draw and label triangles **B** and **C**.   **(2)**

2 marks

3 Enlarge the triangle below with scale factor $-\frac{1}{2}$, using $(0, 1)$ as the centre of enlargement. **(2)**

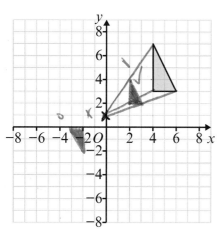

marks

4

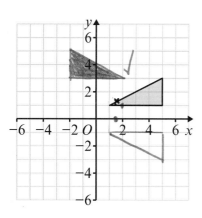

The triangle in this diagram is reflected in the line $y = x$ and then rotated $180°$ about the point $\left(\frac{3}{2}, \frac{3}{2}\right)$

Write down the coordinate(s) of any invariant points under this combination of transformations. **(2)**

stay the same

Answer: ......(2, 1)........................

marks

Make a plan

 Had a go
0–4 marks

Make sure that you understand the vocabulary of transformations. Then you can tackle the core skill of describing transformations.

☑ **Nearly there**
5–7 marks

Well done! Use the hints in the answers to work out where you could have picked up more marks. Remember that if a question asks you to describe a transformation, you must give as much detail as possible.

☑ **Nailed it!**
8–10 marks

Congratulations! Keep an eye out for questions where you have to carry out two or more successive transformations and then describe a single equivalent transformation.

8

MY TOTAL MARKS

Answers
on page 111

Revision Guide pages 104, 105

32 Circle facts and theorems

10

You need to be able to apply and prove standard circle theorems.

My marks

1 Which of the following statements is false?
Tick **one** box. **(1)**

- ☐ **A** Opposite angles in a cyclic quadrilateral add up to 180°
- ☐ **B** The angle in a semicircle is 90°
- ☐ **C** Angles in the same segment add up to 180°
- ☐ **D** The angle between a radius and a tangent is 90°

marks

2 Calculate the angle marked x in this diagram and state the reason for your answer.
O is the centre of the circle. **(2)**

Core skill

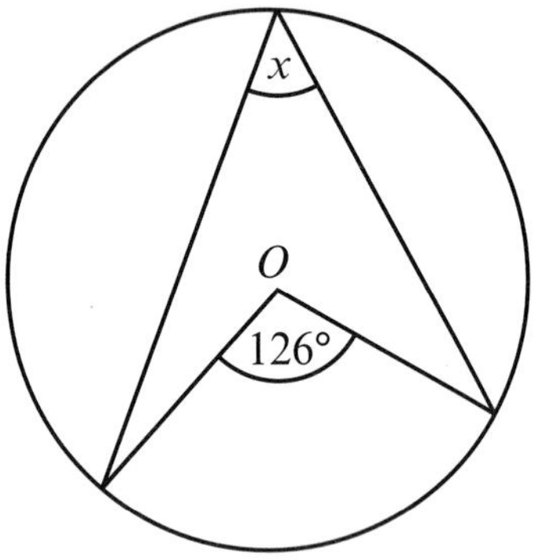

Answer: x = ... °

Reason: ....

marks

3 $ABCD$ is a cyclic quadrilateral.
SBT is a tangent to the circle at B.
Angle SBA = 41° and angle BAC = 37°.

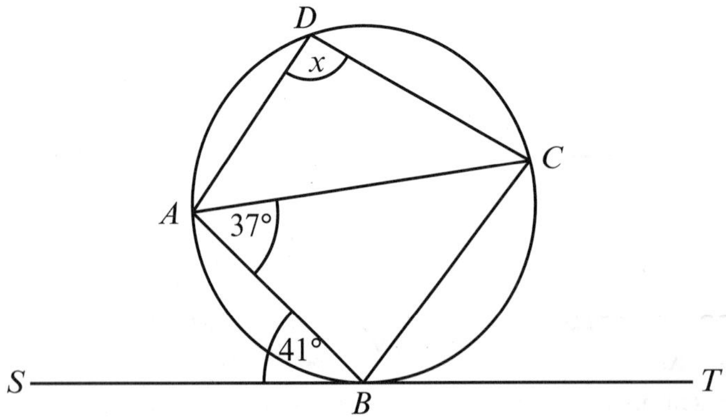

Work out the size of angle x. Give reasons for your answer. **(4)**

x = ...°

....

....

....

4 The diagram shows a circle with centre O, and a diameter AB. Point C lies on the circumference of the circle. Angle $OAC = x$ and angle $OBC = y$.

(a) Write down expressions for angles OCA and OCB in terms of x and y. Give a reason for your answers. **(2)**

Angle $OCA =$ ....   Angle $OCB =$ ....

Reason: ....

(b) Without using any circle theorems, prove that angle $ACB = 90°$. **(3)**

☐

marks

☐

marks

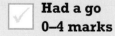

Revision Guide pages 80, 81, 83, 84, 92

33 Area and perimeter

10 You need to be able to use standard units of measure, understand scales, use bearings and know and apply formulae to calculate perimeters and areas.

My marks

1 A triangle has base 9 cm and perpendicular height 6 cm.
Calculate the area of the triangle. **(1)**

Answer: .... cm²

marks

2 A circle has an area of 81π cm².
Calculate the diameter of the circle. **(1)**

Answer: ....cm

marks

3 The sector shown below has a radius of 7 cm and the angle at the centre is 110°.

Core skill

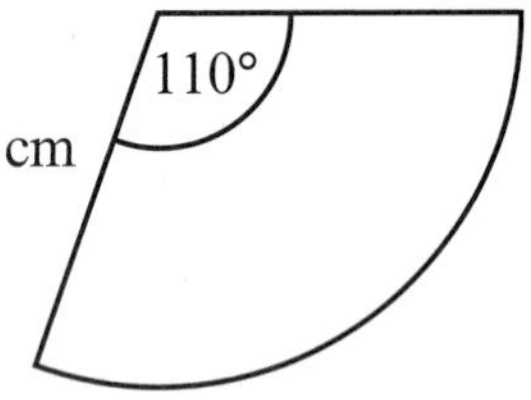

110°

7 cm

Giving your answers to 3 significant figures,

(a) calculate the area of the sector **(1)**

$$\frac{\square}{\square} \times \pi \times \square\square^{\square}$$

Answer: .... cm²

marks

(b) calculate the perimeter of the sector. **(2)**

Answer: .... cm

marks

4 Identify the area equivalent to 10 m².
Tick **one** box. **(1)**

☐ **A** 10 000 cm²
☐ **B** 100 000 cm²
☐ **C** 1 000 000 cm²
☐ **D** 1000 cm²

marks

5 A map has a scale of 1 : 15 000
Calculate, in km, the distance in real life represented by a distance of 9 cm on the map. **(1)**

Answer: .... km

marks

6 The diagram shows a parallelogram joined to a trapezium.

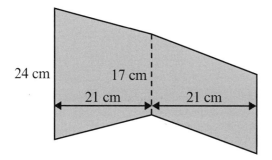

24 cm

17 cm

21 cm   21 cm

Calculate the area of this shape. **(2)**

Answer: .... cm²

marks

7 An ornamental pool is in the shape of a three-quarter circle of radius 3 m.
A paved area surrounds the curved edge of the pool and has a constant width of 0.5 m.
The paved area is to be edged with fencing that costs £11.50 per m.

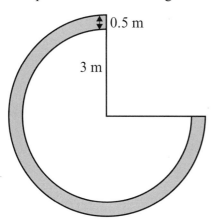

0.5 m

3 m

Calculate the cost of the fencing. **(2)**

Answer: £ .... marks

34 Surface area and volume

  You need to know how to calculate volumes and surface areas of 3-D shapes and apply the concept of similarity to 3-D shapes.

 1 The prism shown below has a cross-section in the shape of a trapezium.

Core skill

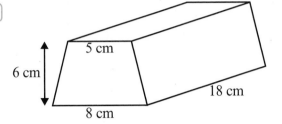

Calculate the volume of the prism. **(2)**

Area of cross-section = ... cm²

Answer: Volume = ... cm³   **marks**

 2 A cylinder has a volume of 180π cm³ and a base radius of 1.5 cm.
Calculate the height of the cylinder. **(2)**

Answer: .... cm   **marks**

 3 Two similar cones have base diameters 18 cm and 30 cm respectively.

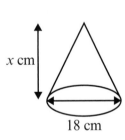 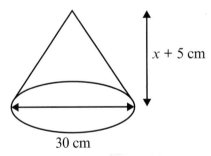

18 cm   30 cm

The height of the smaller cone is x cm and the height of the larger cone is $x + 5$ cm.
Calculate the value of x. **(2)**

Answer: x = ...   **marks**

4 Two similar jars of jam have side lengths in the ratio 1 : 1.5
The surface area of the smaller jar is 320 cm².

Core skill   The volume of the larger jar is 1440 cm³.

   (a) Calculate the surface area of the larger jar. **(1)**

Answer: .... cm²   marks

   (b) Calculate the volume of the smaller jar. **(1)**

Answer: .... cm³   marks

 5 A children's wooden toy is made from a hemisphere of radius 9 cm stuck on top of an inverted cone of height 14 cm.

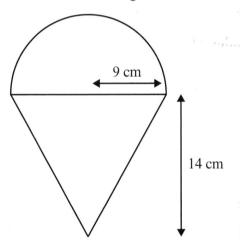

Volume of a sphere $= \frac{4}{3}\pi r^3$

Volume of a cone $= \frac{1}{3}\pi r^2 h$

Calculate the exact volume of the toy. **(3)**

Answer: .... cm³   marks

Make a plan

 Had a go
0–4 marks

Make sure that you are confident with finding the areas of common 2-D shapes and the volumes of cubes and cuboids. Remember that if a question asks you to find the volume of a prism, you need to find the area of the cross-section first.

 Nearly there
5–8 marks

Well done! If you need to use the formulae for the volume of a cone or sphere they will be given to you with the question. Remember that if a questions asks for an exact answer, you should leave your answer in terms of π.

 Nailed it!
9–11 marks

Congratulations! Keep an eye out for questions where you have to deduce the relationship between lengths and/or surface areas of 3-D shapes using algebra.

MY TOTAL MARKS

Answers
on page 113

35 Right-angled triangles

You need to be able to know and use the formulae for Pythagoras' theorem and the trigonometric ratios in 2 and 3 dimensions.

1 Calculate the value of x in this right-angled triangle.

Core skill

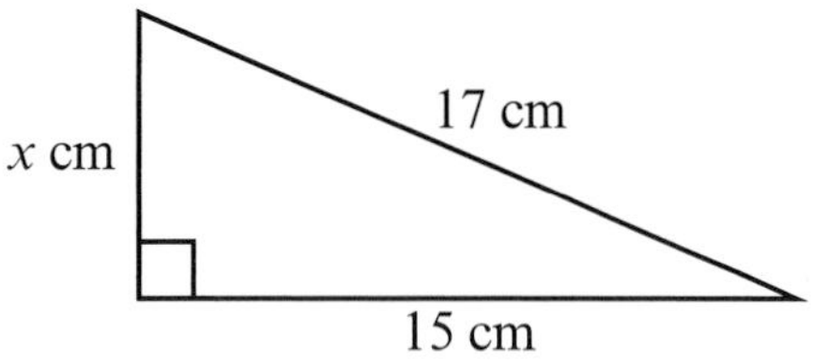

x cm, 17 cm, 15 cm

$17^2 - 15^2 = 64$
$\sqrt{64} = 8$

Answer: $x = $ ....8........ cm

(1) 1 marks

2 Calculate the size of angle x in this right-angled triangle.

Core skill

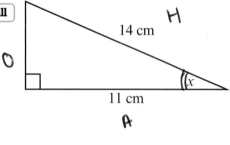

14 cm, O, 11 cm, A, H, x

$CAH = COS$

$= \frac{11}{14} =$

$\cos(x°) = \frac{11}{14}$

$x = \frac{11}{\cos x}$

$\cos^{-1}\left(\frac{11}{14}\right)$
$= 38.21$

Answer: $x = $ .....38.21.....°

(1) 1 marks

3 Calculate the value of x in the diagram below.

7 cm, 5 cm, H, 21°, x cm

Length $a =$
$\sqrt{7^2 - 5^2} = 4.9$

$b = TOA = Tan$

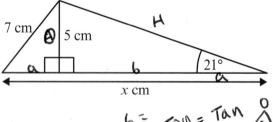

$b = \frac{5}{\tan(21)} = 13.03 + 4.9$

Answer: $x = $ ....17.9....... cm

(3) 3 marks

4 Calculate the angle marked y in this diagram.

y, 25 cm, 40°, 10 cm, O, H

$OAH = Sin \; Cos$

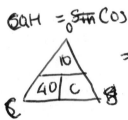

$= \frac{10}{Sin40} = \frac{15.56}{13.1}$
$\cos 40$

$\sin^{-1}\frac{13.65}{25}$
$= 31.5$

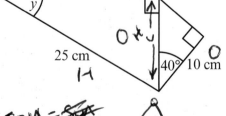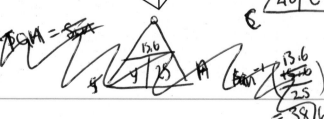

Answer: $y = $ ..31.5........°

(2) 2 marks

5 Jo stands at the top of a clock tower 32 m tall.
The angle of depression to Keith is 35°.
The angle of depression to Lisa is 46°.

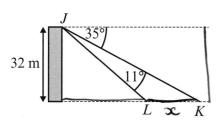

Work out how far apart Lisa and Keith are.

Base of tower to L = $\frac{32}{\tan(46)}$ = 30.90...

to K = $\frac{32}{\tan(35)}$ = 45.70...

(2)

45.70 - 36.90 = 14.8m

Answer: .... m

marks

6 The diagram below shows a cuboid *ABCDEFGH* with dimensions 3 cm, 4 cm and 6 cm.

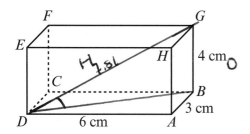

Calculate

$\sqrt{4^2 + 3^2 + 6^2}$ = 7.81 cm

(a) length *GD*

①

Answer: ......7.81....... cm ✓

1 marks

(b) angle *GDB*.

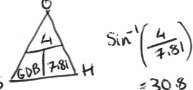

$\sin^{-1}\left(\frac{4}{7.81}\right)$

= 30.8

②

2 marks

Answer: .....30.8......° ✓

Make a plan

MY TOTAL MARKS
10

Answers
on page 114

36 Trigonometry

 You need to know the exact values of the sine, cosine and tangent of some angles. You also need to be able apply the sine rule, the cosine rule and the general area formula for a triangle to solve problems involving non-right-angled triangles.

1 One pair of these stated ratios both have value $\frac{\sqrt{3}}{2}$. Identify which pair.
Tick **one** box. **(1)**

☐ **A** sin 30° and cos 60°
☐ **B** sin 60° and cos 30°
☐ **C** sin 60° and cos 60°
☐ **D** sin 30° and cos 30°

marks

2 The diagram shows a right-angled isosceles triangle.

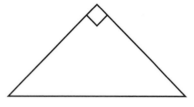

Use the diagram to explain why tan 45° = 1. **(2)**

....

....

marks

3 The cosine rule is...
Tick **one** box. **(1)**

☐ **A** $a^2 = b^2 + c^2 - 2ab\cos C$
☐ **B** $a^2 = b^2 + c^2 + 2bc\cos A$
☐ **C** $a^2 = b^2 + c^2 - 2bc\cos A$
☐ **D** $a^2 = b^2 + c^2 + 2ab\cos B$

marks

4 Work out the value of x in this diagram. **(2)**

Core skill

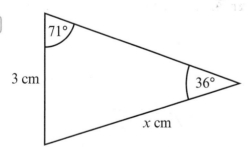

Answer: $x = $ .... cm

marks

5 Calculate the angle marked x in this diagram. (3)

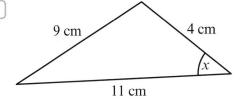

Answer: $x =$ ....°

 marks

6 Calculate the area of the minor segment of this circle with angle $AOB = 80°$. (3)

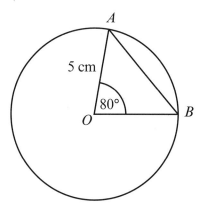

Answer: Area of segment = ... cm² marks

Make a plan

 Had a go
0–4 marks

Make sure you are confident with basic right-angled triangle trigonometry before practising the sine rule and the cosine rule.

 Nearly there
5–9 marks

Well done! Use the hints in the answers to work out where you could have picked up more marks. Always remember to label the sides and angles of a triangle before choosing whether to use the sine or cosine rule.

 Nailed it!
10–12 marks

Congratulations! Keep an eye out for questions where you have to use trigonometric ratios in a non-calculator setting. You should know the exact values of 30°, 45°, 60° and 90° for each of the three trigonometric ratios.

MY TOTAL MARKS

Answers on page 114

Revision Guide pages 106, 107

37 Vumbers

37 Vectors

 You need to be able to describe translations using vectors and apply arithmetical methods to vectors including adding, subtracting and multiplying by a scalar. You also need to be able to construct vector proofs.

1 Write down the vector that maps shape **A** onto shape **B**. **(1)**

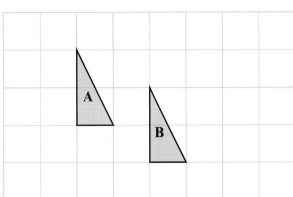

Answer: ....

My marks ⌄

☐ marks

2 $a = \begin{pmatrix} 4 \\ 2 \end{pmatrix}$, $b = \begin{pmatrix} 3 \\ -1 \end{pmatrix}$ and $c = \begin{pmatrix} -2 \\ 6 \end{pmatrix}$

Core skill

Calculate the following vectors.

(a) $a + b$ **(1)**

Answer: .... marks

(b) $c - b$ **(1)**

Answer: .... marks

(c) $4a + 3b - c$ **(2)**

Answer: .... marks

 3 $b = \begin{pmatrix} 8 \\ -3 \end{pmatrix}$ and $c = \begin{pmatrix} -2 \\ 5 \end{pmatrix}$

Find a vector **a** such that $2a + 3b = 4c$ **(2)**

Answer: **a** = ...

☐ marks

4 The diagram shows a parallelogram *ABCD*.
N is the midpoint of *DC* and *M* is the midpoint of *BC*.

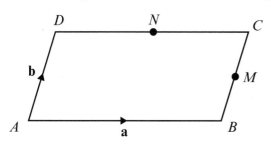

Vector **a** = $\overrightarrow{AB}$ and vector **b** = $\overrightarrow{AD}$.
Show that vectors $\overrightarrow{DB}$ and $\overrightarrow{NM}$ are parallel.

(3) marks

38 Probabilities and outcomes

  You need to understand that experimental probabilities tend towards theoretical probabilities with increasing sample size. You also need to be able to construct possibility spaces for equally likely outcomes and use these to calculate theoretical probabilities.

1 A bag contains red, blue and white balls in the ratio $5:4:3$
Emma picks out a ball at random.
What is the probability the ball is blue? **(1)**

$$P(\text{blue}) = \frac{\boxed{}}{\boxed{}}$$

Answer: .... marks

2 Kate plays a game with two sets of cards.

Set 1 | 1 | 3 | 5 | 7 | 9 |

Set 2 | 2 | 4 | 6 | 8 |

Kate takes one card at random from each set.
She adds the numbers on the two cards to get the total score.

(a) Complete the table to show all the possible scores. **(1)**

		Set 1				
		1	**3**	**5**	**7**	**9**
Set 2	**2**	3	5	7	9	11
	4	5	7	9		
	6					
	8					

marks

(b) What is the probability that Kate's total score will be greater than 12? **(2)**

$$P(\text{greater than 12}) = \frac{\boxed{}}{\boxed{}}$$

Answer: .... marks

3 Four friends each throw a dice a number of times.
The table shows, for each friend, the number of trials and the number of times the dice landed on 6.

 Core skill

	Ali	Ben	Caz	Dom
Number of trials	20	40	100	60
Number of 6s	3	8	25	12

(a) Use all the results in the table to work out an estimate for the probability that the dice will land on 6. **(2)**

Estimated probability = $\dfrac{\boxed{}}{\boxed{}}$

Answer: ....

marks

(b) Write down the theoretical probability of getting a 6 on a fair dice. **(1)**

Answer: ....

marks

(c) Assuming the dice is fair, complete the following statement. **(1)**

....................'s results will give the best estimate for the theoretical probability.

Because they show the .... number of ....

marks

4 The table shows the probabilities that a biased dice will land on 1, 2, 3 or 4.
The probability the dice will land on 6 is twice the probability it will land on 5.

Number on dice	1	2	3	4	5	6
Probability	0.14	0.08	0.16	0.11	x	$2x$

(a) Work out the probability the dice will land on 6. **(3)**
You must show your working.

.... + ... + ... + ... + ... + ... = 1

Answer: ....

marks

(b) Bradley rolls the biased dice 200 times.
Work out an estimate for the total number of times the dice will land on 3. **(1)**

Answer: ....

marks

39 Venn diagrams

  You need to be able to draw and interpret Venn diagrams, including the use of set notation, and use them to calculate the probability of independent combined events.

1 In a group of 60 students surveyed:

40 have a sister

35 have a brother

7 have neither a brother nor a sister.

Complete the Venn diagram for this information. **(3)**

[ ] marks

2 Here is a Venn diagram.

Core skill

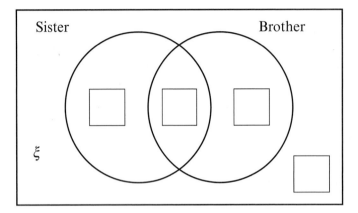

(a) How many numbers are in the universal set ξ? **(1)**

Answer: .... marks

(b) Write down the numbers that are in these sets:

(i) $A \cup B$ **(1)**

Answer: $A \cup B = $ .... marks

(ii) $A \cap B$ **(1)**

Answer: $A \cap B = $ .... marks

One of the numbers in the diagram is chosen at random.

(c) Find the probability that the number is in set A' (2)

$$P \text{ (number is in set } A') = \frac{\boxed{}}{\boxed{}}$$

Answer: ....

marks

3 In a set of numbers:
ξ = odd numbers between 0 and 24
A = 1, 7, 9, 13, 17
B = 1, 5, 7, 19
C = 7, 17, 19, 21

(a) Complete the Venn diagram for this information. (2)

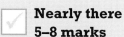

A number is chosen at random from ξ

marks

(b) Find the probability that the number is a member of $A \cap B'$. (2)

Answer: ....

marks

Make a plan

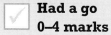 **Had a go**
0–4 marks
Venn diagrams are a useful way of sorting data. Make sure you feel confident working with Venn diagrams and set notation.

☑ **Nearly there**
5–8 marks
Well done! Use the hints in the answers to work out where you could have picked up more marks. You need to learn the meaning of the symbols used in set notation to be able to answer questions based on Venn diagrams.

☑ **Nailed it!**
9–12 marks
Congratulations! One useful quick check for probability answers is that they are always decimals or fractions between 0 and 1.

 MY TOTAL MARKS

Answers on page 116

40 Independent events and tree diagrams

10 You need to be able to calculate the probability of independent combined events, and use tree diagrams to solve problems.

1 Event A and event B are independent events.
The probability that event A will happen is 0.4
The probability that event B will happen is 0.7

(a) Complete the probability tree diagram. **(2)**

Event A Event B

☐ will happen
will happen
☐ will not happen

☐ will happen
will not happen
☐ will not happen

☐ marks

(b) Work out the probability that either event A will happen or event B will happen but not both.
You must show your working. **(2)**

P(A will happen and B will not happen) = ☐ × ☐
P(A will not happen and B will happen) = ☐ × ☐

P(A will happen and B will not happen) = ...

P(A will not happen and B will happen) = ...

Answer: ....

☐ marks

2 Abi plays a game against Matt.
Abi is three times as likely as Matt to win the game.

Core skill The probability that the game is drawn is 0.2

(a) Work out the probability that Matt will win the game. **(2)**

Answer: ....

☐ marks

(b) Work out the probability that Matt wins at least one out of three games. **(2)**

Answer: ....

☐ marks

3 Becca throws a fair dice until it lands on 6.

Work out the probability it will take 4 or more throws of the dice before it lands on 6.

You must show your working. **(4)**

P(takes 1 throw) = $\dfrac{\square}{\square}$

P(takes 2 throws) = $\dfrac{\square}{\square} \times \dfrac{\square}{\square}$

P(takes 3 throws) =

Answer: ....

marks

Make a plan

✓ **Had a go**
0–4 marks

Make sure you are confident calculating probabilities of single events before you try to answer questions about combined events.

✓ **Nearly there**
5–8 marks

Well done! Use the hints in the answers to work out where you could have picked up more marks.

You can often use tree diagrams to solve problems involving more than one event.

✓ **Nailed it!**
9–12 marks

Congratulations! Being confident with this material will help you tackle conditional probability problems.

MY TOTAL MARKS

Answers on page 117

41 Conditional probability

   You need to be able to use two-way tables, tree diagrams and Venn diagrams to tackle conditional probability questions.

1 If Adam goes to bed before midnight on a school night, the probability he will sleep through his alarm is 0.1.
If he goes to bed at midnight or later the probability he will sleep through his alarm is 0.7.
The probability Adam goes to bed before midnight on a school night is 0.2.

(a) Complete the probability tree diagram for these events. **(2)**

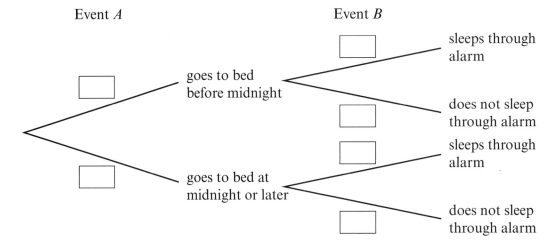

Event *A*   Event *B*

goes to bed before midnight → sleeps through alarm / does not sleep through alarm

goes to bed at midnight or later → sleeps through alarm / does not sleep through alarm

marks ☐

(b) Work out the probability Adam sleeps through his alarm on a school night.
You must show your working. **(2)**

Answer: ....   marks ☐

 2 The table shows where students went on a school trip.

Core skill

	London	Edinburgh	Cardiff
Male	15	18	12
Female	24	16	14

(a) Given that a student went to Edinburgh, what is the probability that they are female? **(2)**

Number of students who went to Edinburgh = ...

Answer: ....   marks ☐

(b) Given that a student is male, what is the probability they went to Cardiff? **(2)**

Number of male students = ...

Answer: ....   marks ☐

3 The Venn diagram shows the sports activities of a group of 100 students.

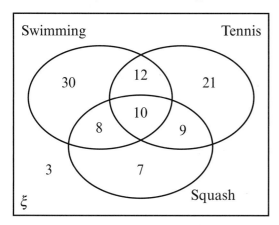

(a) A student is picked at random.
Given that this student plays tennis, work out the probability they also go swimming.
You must show your working. **(2)**

Number of students that play tennis = ...

Answer: ....

marks

(b) Another student is picked at random.
Given that this student takes part in at least two activities, work out the probability they play squash.
You must show your working. **(2)**

Number of students that take part in at least two activities = ...

Answer: ....

marks

Make a plan

 **Had a go
0–4 marks**

Conditional probability is tricky. Think about restricting the sample space using the conditions given in the question.

 **Nearly there
5–8 marks**

Well done! Use the hints in the answers to work out where you could have picked up more marks. Make sure probability answers are decimals or fractions between 0 and 1.

 **Nailed it!
9–12 marks**

Congratulations! You don't need to simplify fractions in probability answers unless you are told to do so. And it's easier to check your working if you don't simplify.

 MY TOTAL MARKS

Answers
on page 119

42 Sampling, averages and range

  You need to know about sampling methods and calculate averages and measures of spread from discrete and continuous data.

1 The table shows the masses, in kg, of 50 cats.

Core skill

Mass (x kg)	Frequency
$3 \leqslant x < 4$	21
$4 \leqslant x < 5$	17
$5 \leqslant x < 6$	7
$6 \leqslant x < 7$	5

Calculate an estimate for the mean mass of the cats. **(3)**

Mean mass = ... kg

2 The stem-and-leaf diagram shows the ages of people on a train.

```
0 | 9  9
1 | 5  6  8  9  9
2 | 1  1  2  2  4  5  7  8
3 | 0  1  4  7  8  8
4 | 0  1  4  6
```

Key: 1|5 means 15 years of age.

(a) Write down the median age and the range of the ages. **(2)**

Answer: Median = ...

Answer: Range = ...

(b) Find the interquartile range. **(2)**

Answer: IQR = ...

The ages of people on a bus had a median of 40 and an interquartile range of 26.

(c) Compare the ages of the people on the train and on the bus. **(1)**

....

....

....

3 Ambika wants to find out about the types of TV programmes watched by the people in her school.
She asks five people from her class.

(a) State two ways in which Ambika can improve her sampling to reduce the likelihood of bias. **(1)**

1 ....

2 ....

marks

She decides to take a random sample of size 40.

(b) Suggest one method she could use to select a random sample. **(1)**

....

marks

4 Phoebe wants to find an estimate of the number of deer in a forest.
She catches a sample of 50 deer and tags each of them.
These deer are then released back into the forest.
The next week she catches a sample of 40 deer and finds that 16 of them are tagged.
Work out an estimate for the total number of deer in the forest. **(2)**

Answer: .... deer

marks

Make a plan

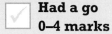 **Had a go**
0–4 marks
Make sure you know how to work out the mean and median of discrete data. Then you can tackle the core skill of calculating the mean from a table of grouped continuous data.

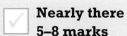 **Nearly there**
5–8 marks
Well done! Remember that if a question asks you to compare two sets of data, you should compare both the average and the spread.

Nailed it!
9–12 marks
Congratulations! Keep an eye out for questions where you have to calculate the mean from a data set displayed as a graph.

 MY TOTAL MARKS

Answers on page 120

43 Representing data

  You need to be able to use appropriate graphical representations such as cumulative frequency diagrams, box plots and scatter diagrams.

1 The cumulative frequency diagram shows the lengths, in cm, of 120 worms.

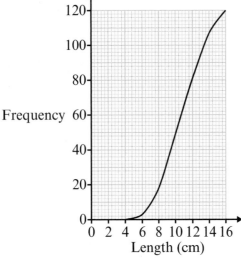

(a) Estimate the median length of the worms. **(1)**

Answer: Median = ... cm

(b) Estimate the interquartile range of the lengths of the worms. **(2)**

Answer: IQR = ...

(c) Draw a box plot to display this data. **(2)**

2 The scatter graph shows the temperature, $x°C$, and the number of hours of sunshine, y hours.

Core skill

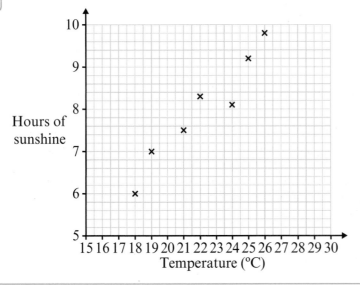

My marks

Two more data values are recorded.

There were 8.6 hours of sunshine on a day when the temperature was 24°C and there were 7 hours of sunshine on a day when the temperature was 21°C.

(a) Plot these two points on the scatter graph. (1)

marks

(b) Draw a line of best fit on the scatter graph. (1)

(c) Describe the type of correlation shown. (1)

marks

Answer: ....

(d) Use your line of best fit to estimate the number of hours of sunshine when the temperature is 20°C. (1)

marks

Answer: .... hours

marks

(e) State, with a reason, whether this estimate is likely to be reliable. (1)

....

marks

3 The scatter graph shows the sales of ice creams, x '000s, and the sales of sun cream, y '000s.

Harpreet uses the scatter graph to predict the sales of sun cream when there are 11 000 ice creams sold.

Give two reasons why Harpreet should not do this. (2)

1 ....

2 ....

marks

Make a plan

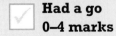 **Had a go**
0–4 marks

You need to make sure that you understand what is meant by a median or quartile value. Then you can tackle questions where you have to read these values from a graph.

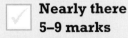 **Nearly there**
5–9 marks

Well done! Remember that if a question asks you to comment on the reliability of an estimate, you should look at whether the prediction is within the range of the data.

Nailed it!
10–12 marks

Congratulations! Keep an eye out for questions where you have to compare data sets using either cumulative frequency diagrams or box plots.

MY TOTAL MARKS

 Answers
on page 120

44 Representing data (continued)

  You need to be able to use and interpret time series graphs, histograms and frequency polygons.

 1 The table shows the monthly rainfall in a town over a 6-month period.

Month	April	May	June	July	August	September
Rainfall (mm)	21	15	12	9	10	18

(a) Draw a time series graph to represent this data. **(2)**

marks

(b) Describe the trend. **(1)**

....

marks

2 The table shows the masses, m kg, of a group of gorillas.

Core skill

Mass (m kg)	$135 \leqslant m < 150$	$150 \leqslant m < 160$	$160 \leqslant m < 165$	$165 \leqslant m < 170$	$170 \leqslant m < 190$
Frequency	30	85	90	75	40

(a) Draw a histogram to display this data. **(2)**

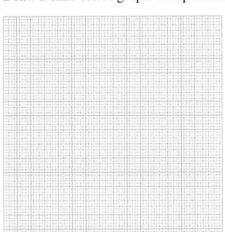

(b) On the same axes, draw a frequency polygon. **(2)**

marks

3 The histogram shows the times taken by 80 runners in a 400 m race.

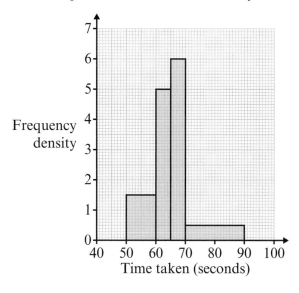

(a) Explain why a histogram has been used to display this data. **(1)**

....

.... marks

(b) Calculate an estimate for the number of runners who took between 60 and 80 seconds to complete the race. **(2)**

Answer: .... marks

Answers

Number

1 Fractions and decimals

1 **D** $\underline{0.45}$ ✔

$$\frac{9}{20} \times \frac{5}{5} = \frac{45}{100} = 0.45$$

2 **D** $\underline{61\%}, \underline{\frac{13}{20}}, \underline{\frac{2}{3}}, \underline{0.67}$ ✔

> Convert all to decimals to make it easier to compare.
> $61\% = 0.61, \frac{13}{20} = 0.65, \frac{2}{3} = 0.666...$

3 **(a)** **C** $\underline{\frac{9}{40}}$ ✔

> $\frac{9}{40} = 0.225$

(b) A fraction can be written as a terminating decimal if the prime factors of the $\underline{\text{denominator}}$ ✔ of the fraction only consist of powers of 2 and / or $\underline{5}$. ✔

> **Marking**
> Score 1 mark only if you have both entries correct.

4 $2\frac{3}{5} + 3\frac{1}{2}$

$2 + 3 = 5$

$$\frac{3}{5} + \frac{1}{2} = \frac{\boxed{6}}{10} + \frac{\boxed{5}}{10} = \frac{\boxed{11}}{10} = 1\frac{1}{10}$$ ✔

$$5 + 1\frac{1}{10} = 6\frac{1}{10} \text{ or } \frac{61}{10}$$ ✔

> **Core skill**
> Add two mixed numbers by adding the whole numbers first and then the fractions.

5 $5\frac{1}{4} \div 1\frac{7}{8}$

$$\frac{21}{4} \div \frac{15}{8} = \frac{21}{4} \times \frac{8}{15}$$ ✔

$$\frac{7}{1} \times \frac{2}{5} = \frac{14}{5} = 2\frac{4}{5}$$ ✔

> **Core skill**
> Convert mixed numbers to improper fractions first. For division, turn the second fraction 'upside down' and change the ÷ to a ×. Check for any factors that will cancel so that you can multiply smaller numbers.

6 $1 - \frac{5}{8} - \frac{1}{5}$ ✔

$$= \frac{40 - 25 - 8}{40} = \frac{7}{40}$$ ✔

> Total area = Area **A** + Area **B** + Area **C**
> Area **B** = 1 − Area **A** − Area **C**

> **Marking**
> Give yourself 1 mark for $1 - \frac{5}{8} - \frac{1}{5}$ and the 2nd mark for the complete answer correct.

7 $n = 0.636363...$

$100n = 63.636363...$ ✔

$99n = 63$

$$n = \frac{63}{99} = \frac{7}{11}$$ ✔

> Multiply by: 10 if 1 digit recurs, 100 if 2 digits recur, 1000 if 3 digits recur.

2 Factors, multiples and counting

1   1, 2, 3, 6, 9, 18, 27, 54 ✔ ✔

To find all the factors of a number identify all the factor pairs:
$1 \times 54, 2 \times 27, 3 \times 18, 6 \times 9$
Remember to include 1 and the number itself.

2 $2^2 \times 3 \times 7$ ✔

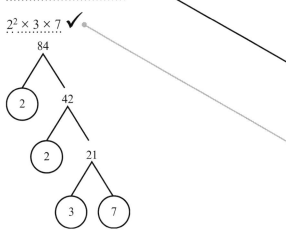

Complete the factor tree by using factor pairs of each number, circling the prime factors as you go along. Continue until every branch ends with a prime number. Write down all the circled numbers putting in multiplication signs, and express repeated multiples as powers: $2 \times 2 = 2^2$

3   As a product of prime factors $180 = 2^2 \times 3^2 \times 5$ ✔

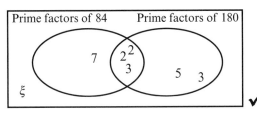

HCF = $2 \times 2 \times 3 = 12$ ✔

The HCF of two numbers is the largest number that is a factor of both numbers.
To use a Venn diagram to find the HCF, write the numbers as a product of their prime factors and put the common factors in the intersection. The HCF is the product of all the prime factors in the intersection.

4

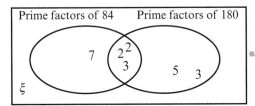

LCM = $2 \times 2 \times 3 \times 7 \times 3 \times 5 = 1260$ ✔

The LCM is the product of all the prime factors in the Venn diagram.

5 $4 \times 3 \times 2 \times 1 = 24$ ✔

6   There are 26 letters in the alphabet, 4 possible even non-zero numbers and 5 possible odd numbers.

Total number of possible passwords

= $\boxed{26} \times \boxed{4} \times \boxed{5}$ ✔

= 520 ✔

3 Powers and roots

1 B $\underline{2^5}$ ✔

$2^5 = 2 \times 2 \times 2 \times 2 \times 2 = 32$

2 (a) $\sqrt{225} = \underline{15}$ ✔

You need to be able to remember the square numbers and their associated square roots up to 15^2

(b) $5^3 = 5 \times 5 \times 5 = \underline{125}$ ✔

125 is a cube number – a number multiplied by itself then multiplied by itself again. You need to be able to remember the cubes of 2, 3, 4, 5 and 10.

3 (a) $(3^2)^3 \times 3^0 = 3^{6+0} = \underline{3^6}$ ✔

(b) $\dfrac{3^4 \times 3^2}{3} = \dfrac{3^{4+2}}{3^1} = 3^{6-1} = \underline{3^5}$ ✔

Using the laws of indices
$a^m \times a^n = a^{m+n}$ and $(a^m)^n = a^{mn}$

4 B $\underline{13.4}$ ✔

Using the laws of indices $a^m \times a^n = a^{m+n}$
and $a^m \div a^n = a^{m-n}$

5 $x^{-\frac{3}{2}} = \dfrac{27}{125}$

$x^{\frac{3}{2}} = \dfrac{\boxed{125}}{\boxed{27}}$ ✔

$13^2 = 169$ and $14^2 = 196$, so $\sqrt{180}$ is between 13 and 14

$x^{\frac{1}{2}} = \left(\dfrac{125}{27}\right)^{\frac{1}{3}} = \dfrac{5}{3}$ ✔

$x = \left(\dfrac{5}{3}\right)^2 = \underline{\dfrac{25}{9}}$ ✔

Core skill

Use $\left(\dfrac{a}{b}\right)^{-n} = \left(\dfrac{b}{a}\right)^n$ to write $x^{\frac{3}{2}} = \dfrac{125}{27}$
Take the cube root of both sides to find $x^{\frac{1}{2}}$
Then square both sides. Remember to give your answer in exact form.

6 $2^{3x-5} = \dfrac{1}{8}$

$2^{3x-5} = 2^{-3}$ ✔

$3x - 5 = -3$ ✔

$3x = 2$

$x = \underline{\dfrac{2}{3}}$ ✔

Start by expressing the right-hand side of the equation as a power of 2, then equate the powers and solve for x.

Look for a factor of 75 that is a square number. Use the rule $\sqrt{ab} = \sqrt{a} \times \sqrt{b}$ to split the square root into two square roots then write $\sqrt{25}$ as a whole number.

4 Exact answers and standard form

Rationalising the denominator means making the denominator a whole number. You can do this by multiplying the top and bottom of the fraction by the surd part in the denominator. Remember that $\sqrt{3} \times \sqrt{3} = 3$

1 $\sqrt{75} = \sqrt{25 \times 3} = \sqrt{25} \times \sqrt{3} = \underline{5\sqrt{3}}$ ✔

2 $\dfrac{7}{2\sqrt{3}} = \dfrac{7}{2\sqrt{3}} \times \dfrac{\sqrt{3}}{\sqrt{3}} = \dfrac{7\sqrt{3}}{2\sqrt{3} \times \sqrt{3}} = \dfrac{7\sqrt{3}}{\underline{6}}$ ✔

Marking

Give yourself 1 mark for correctly writing
as $\sqrt{2 \times 36} + \sqrt{2 \times 121}$

3 $\sqrt{72} + \sqrt{242} = \underline{\sqrt{2 \times 36} + \sqrt{2 \times 121}}$ ✔

$= \sqrt{2} \times \sqrt{36} + \sqrt{2} \times \sqrt{121} = 6\sqrt{2} + 11\sqrt{2} = 17\sqrt{2}$

so $k = \underline{17}$ ✔

You want an answer in the form $k\sqrt{2}$ so divide 72 and 242 by 2 to find the square factor.

4 $6.7 \times 10^6 = \underline{6\,700\,000}$
and $3.4 \times 10^5 = \underline{340\,000}$ ✔

Marking

Give yourself 1 mark for correctly converting both numbers to ordinary numbers.

$\begin{array}{r} 6\,700\,000 \\ +\ 340\,000 \\ \hline 7\,040\,000 \end{array}$

Core skill

Add or subtract standard form numbers by writing them as ordinary numbers.

$= \underline{7.04 \times 10^6}$ ✔

5 $(1.4 \times 10^{-5}) \div (2 \times 10^{-2})$

$= (1.4 \div 2) \times (10^{-5} \div 10^{-2})$ ✔

Marking

You could also write this as
$\dfrac{1.4}{2} \times \dfrac{10^{-5}}{10^{-2}}$

$= 0.7 \times 10^{(-5-(-2))}$

Core skill

Multiply or divide the number parts and apply the rules of indices to the powers of 10. Rewrite your answer in standard form if necessary.

$= 0.7 \times 10^{-3} = \underline{7 \times 10^{-4}}$ ✔

6 $\dfrac{10}{3}\pi - \dfrac{12}{5}\pi = \left(\dfrac{\boxed{50} - \boxed{36}}{15}\right)\pi$ ✔

$\qquad\qquad = \dfrac{14}{15}\pi \,\mathrm{cm}$ ✔

Read the question carefully. You need to give an exact answer in terms of π.

7 $36\pi \div \dfrac{16}{9}\pi = \dfrac{36\pi}{1} \div \dfrac{16\pi}{9}$

$\qquad\qquad = \dfrac{36\pi}{1} \times \dfrac{9}{16\pi}$ ✔ ○

$\qquad\qquad = \dfrac{9\pi}{1} \times \dfrac{9}{4\pi}$

$\qquad\qquad = \dfrac{81}{4}$ or 20.25 ✔

Marking

Give yourself 1 mark if you turned the second fraction upside down and changed $\div$ to $\times$.

$\dfrac{16}{9}\pi = \dfrac{16\pi}{9}$, so you can write the calculation as $\dfrac{36\pi}{1} \div \dfrac{16\pi}{9}$.

Check to see if there are any factors that will cancel so that you can multiply smaller numbers. In this question π will cancel.

5 Calculator skills

You can count decimal places to convert between ordinary numbers and standard form.
The power of 10 is negative so the number is < 1

1 $7.6 \times 10^{-4} = 0.00076$ ✔

2 $580\,000\,000 = 5.8 \times 10^8$ ✔

3 $(3.6 - 0.55)^2 = 9.3025$

$\sqrt[3]{10.648} = 2.2$ ✔ ○

$9.3025 + 2.2 = 11.5025$ ✔

Marking

Give yourself 1 mark for one of the two initial calculations correct.

Marking

Give yourself 1 mark for one of the two initial calculations correct.

4 $\sqrt{12.5 + 3.4} = 3.987\,480\,407$

$4.2^3 = 74.088$ ✔ ○

$3.987\,480\,407 \div 74.088$
$= 0.053\,820\,8\,67(18\ldots)$ ✔ ○

Work out the numerator (top) and denominator (bottom) separately and write them both down before dividing.
You are asked to write down all the figures on your calculator display so do not round your answer.

Marking

You can get the 2nd mark for the correct answer provided you have all the digits up to the bracket.

5 $(3.72 \times 10^{-4}) \times (2.1 \times 10^7)$
$= 7812 = 7.812 \times 10^3$ ✔ ✔ ○

Marking

Give yourself 1 mark for 7.812 and 1 mark for $\times 10^3$

6 $\dfrac{2.625 \times 10^5}{5.25 \times 10^{-3}} = (2.625 \times 10^5) \div (5.25 \times 10^{-3})$

$= 50\,000\,000 = 5 \times 10^7$ ✔ ✔ ○

Core skill

Remember, for the answer to be in standard form $A \times 10^n$, then $1 \leqslant A < 10$ and n is an integer.

7 $\text{mean} = \dfrac{6.69 \times 10^7}{2.425 \times 10^5}$ or

$(6.69 \times 10^7) \div (2.425 \times 10^5)$ ✔

$= 275.87$

275 people ✔

Marking

Give yourself 1 mark for 5 and 1 mark for $\times 10^7$

Core skill

You can do this question on your calculator using either the division function or the fraction function.

The mean number of people per $\mathrm{km}^2 = \dfrac{\text{population}}{\text{area}}$
The answer should be a whole number, because it represents the number of people.

6 Estimation and accuracy

1 $5.23 \approx 5$

$3.47 \approx 3$

$0.472 \approx 0.5$ ✔ ○

$\dfrac{5.23 \times 3.47}{0.472} \approx \dfrac{5 \times 3}{0.5}$

$\qquad = \dfrac{15}{0.5} = 30$ ✔

Marking

Give yourself 1 mark if two out of the three roundings are correct.

For estimation questions, round each number in the calculation to 1 significant figure.

2 **(a)** Lower bound for $18\,\text{cm} = \boxed{17.5}\,\text{cm}$

Lower bound for $11\,\text{cm} = \boxed{10.5}\,\text{cm}$ ✔

Lower bound for perimeter

$= 2(17.5 + 10.5) = 56\,\text{cm}$ ✔

The perimeter will have its lower bound when both lengths have their lower bound.

(b) Lower bound for area $= 17.5 \times 10.5$

The area will also have its lower bound when both lengths have their lower bound.

$= 183.75\,\text{cm}^2$ ✔

3

	Lower bound	Upper bound
Area of rectangle	25.5	26.5
Length of rectangle	5.05	5.15

✔

To make the width as large as possible, the area (numerator) needs to be as large as possible and the length (denominator) needs to be as small as possible.

$$w = \frac{\text{area}}{\text{length}}$$

Upper bound for width $= \dfrac{\text{upper bound of area}}{\text{lower bound of length}}$

$= \dfrac{26.5}{5.05}$ ✔ $= 5.2475\ldots\,\text{cm}$ ✔

4

	Lower bound	Upper bound
Surface area	119.5	120.5
Radius	3.05	3.15

✔

Curved surface area, $A = 2\pi rh$ so $h = \dfrac{A}{2\pi r}$

Upper bound for height $= \dfrac{120.5}{2 \times \pi \times 3.05}$

$= 6.287\,924\,801\ldots$

Lower bound for height $= \dfrac{119.5}{2 \times \pi \times 3.15}$ ✔

$= 6.037\,782\,762\ldots$

Height $= 6\,\text{cm}$ (1 s.f.) ✔

5 Least possible value $= 5.3$

Greatest possible value is 5.4

The number has been truncated.

Error interval is $5.3 \leqslant n < 5.4$ ✔

Take the highest common factor of the two terms outside of the bracket. $2x$ is common to both terms.

Algebra

7 Brackets and factorising

1 $4x^2 - 6x = 2x(2x - 3)$ ✔ ✔

Make sure you multiply **all** the terms inside the brackets by the term in front of the bracket.

2 $3(x + 7) + 5(2x - 3) = 3x + 21 + 10x - 15$ ✔

$= 13x + 6$ ✔

To expand the brackets, remember FOIL: **F**irst terms, **O**uter terms, **I**nner terms, **L**ast terms.

3 $(x + 4)(x + 5) = x^2 + 5x + 4x + 20$

$= x^2 + 9x + 20$ ✔

4 **B** $(x-7)(x+2)$ ✓

$-14 = -7 \times 2$ and $-5 = -7 + 2$

Core skill

Look for two numbers that multiply together to give the number at the end and add together to give the number in front of the x term.

5 $x^2 - 64 = (x-8)(x+8)$ ✓

This is the difference of two squares.

6 $(2x-3)(3x+4) = 6x^2 + 8x - 9x - 12$

$= 6x^2 - x - 12$ ✓

Use FOIL. Make sure you simplify your answer.

7 $(x-2)(x+4)(x+3) = (x^2 + 2x - 8)(x+3)$ ✓

$= x^3 + 3x^2 + 2x^2 + 6x - 8x - 24$

$= x^3 + 5x^2 - 2x - 24$ ✓

$2x \times x = 2x^2$ and $-1 \times 3 = -3$

8 **A** $(2x-1)(x+3)$ ✓

Core skill

Work out how to split the ax^2 term and the c term and then try expanding some of the options.

9 **D** $(3x+2)(4x-3)$ ✓

$3x \times 4x = 12x^2$ and $-3 \times 2 = -6$. Check that expanding the brackets gives the correct middle term.
$3x \times (-3) + 2 \times 4x = -9x + 8x = -x$

8 Algebraic manipulation

1 250 °C ✓

2 $E = mc^2$

$\frac{E}{m} = c^2$ ✓

$c = \sqrt{\frac{E}{m}}$ ✓

Core skill

Use inverse operations to rearrange formulae.

3 **D** $b = \frac{A-c}{2}$ ✓

In the incorrect answer, the c term should be positive.

4 $\frac{5 \times 6 \times x^{7+6} y^{2-1}}{3x^4 y^{-2}} = \frac{30 \times x^{13} y^1}{3x^4 y^{-2}}$

$= 10x^{13-4} y^{1+2}$

$= 10x^9 y^3$ ✓ ✓

Use the rules of indices to simplify the x terms and the y terms. Don't forget to simplify the numbers as well.

Marking

Score 1 mark if you got two of the three terms correct.

5 1.47 ✓

Take care when you enter the substitutions into your calculator and use brackets.
Don't forget to round your answer.

6 $a(x-y) = 4(b-x)$

$ax - ay = 4b - 4x$

$ax + 4x = 4b + ay$ ✓

$x(a+4) = 4b + ay$

$x = \frac{4b + ay}{a+4}$ ✓

7 $\frac{(x-1)(x-2)}{(x+4)(x-2)}$

C $\frac{x-1}{x+4}$ ✓

Factorise both expressions before cancelling complete brackets.

8 $\frac{2(b+2) + 5 \times 3}{24b}$ ✓

$\frac{2b+19}{24b}$ ✓

9 Functions and proof

1 $3(x + 2) = 3x + 6$ is an example of an <u>identity</u> ✔

> An identity is true for all values of the variable.

2 **A** $\underline{2n + 3}$ ✔

> The next odd number is 2 more than $2n + 1$ so $2n + 1 + 2 = 2n + 3$

3 **(a)** $\underline{13}$ ✔

(b) $\underline{x = 5}$ ✔

4 **B** $\underline{23}$ ✔

> **Core skill**
> Substitute a into the expression for the given function. To solve $f(a) = b$, set up and solve an equation with the function equal to b.

5 $y = 5x + 3$

$5x = y - 3$

$x = \dfrac{y - 3}{5}$

D $f^{-1}(x) = \dfrac{x - 3}{5}$ ✔

> Substitute the value given into g first. $g(2) = 8$ so $f(8) = 3 \times 8 - 1 = 23$

> **Core skill**
> Let $y = f(x)$, rearrange to make x the subject, then swap the xs and ys.
> Give your final answer in the form $f^{-1}(x) =$

6 $2n + (2n + 2) + (2n + 4) = \underline{6n + 6}$ ✔ ✔

$= \underline{6(n + 1)}$ ✔

> **Marking**
> Score 1 mark for each correct gap filled in. You do not need brackets around the $2n + 2$ in the first line.

7 $(n + 1)^2 - (n)^2 = \underline{n^2 + 2n + 1 - n^2}$ ✔

$= \underline{2n + 1}$ ✔

Hence the difference of the squares of two

consecutive integers is always odd. ✔

> **Marking**
> Score 1 mark for writing the two consecutive integers in the correct form and correct order.

10 Straight-line graphs

1 Gradient: $\underline{3}$

y-intercept: $\underline{(0, -2)}$ ✔

> When the equation is given in the form $y = mx + c$, the gradient is m and the y-intercept is the coordinate point $(0, c)$.

2 $2y = -3x + 6$

$y = -\dfrac{3}{2}x + 3$

D $-\dfrac{3}{2}$ ✔

> Rearrange the equation.

3 **(a)** $15 \div 10 =$

$\underline{1.5}$ ✔

(b) $\underline{\text{Cost per mile}}$ ✔

> To interpret the value of the gradient, think about what it means in the context. As the distance increases by 1 mile, the cost increases by £1.50.

4 Gradient = 0.5

y-intercept = 2

Equation: $y = 0.5x + 2$ ✔ ✔

> **Marking**
> Score 1 mark for the correct gradient and 1 mark for the correct y-intercept.

5 $m = \dfrac{8 - 3}{3 - 1} = \dfrac{5}{2}$

$3 = \dfrac{5}{2} \times 1 + c$

$c = 3 - \dfrac{5}{2} = \dfrac{1}{2}$

Equation: $y = \dfrac{5}{2}x + \dfrac{1}{2}$ ✔ ✔

> **Core skill**
> Find the gradient using the two points. Substitute one pair of (x, y) values into $y = mx + c$ and solve to find c.

> **Marking**
> Score 1 mark for the correct gradient and 1 mark for the correct y-intercept. You can use decimals $y = 2.5x + 0.5$ if you prefer.

6 $m = 2$

$3 = 2 \times 1 + c$

$c = 3 - 2$

Equation: $y = 2x + 1$ ✔✔

The product of perpendicular gradients is -1. You find the gradient of a perpendicular line by taking the negative reciprocal.

To find the y-intercept, substitute the given coordinate point into $y = 2x + c$ and calculate the value of c.

Marking

Score 1 mark for the correct gradient and 1 mark for the correct y-intercept.

11 Functions and graphs

1 (a) Missing values: -4 and 2 ✔

(b)

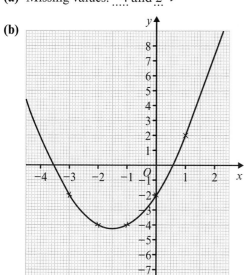

✔

Make sure you join the points with a smooth curve.

Core skill

Read the values from the graph and give them exactly if possible, or rounded to 1 decimal place.

2 (a) $(0, 2)$ ✔

The y-intercept is at the point where $x = 0$

(b) $x = 0.4$

$x = 4.6$ ✔

Marking

Both crossing points need to be correct to get the mark.

The graph crosses the x-axis at the point where $y = 0$

(c) $(2.5, -4.2)$ ✔

Marking

Both values must be correct to 1 decimal place to score the mark.

3 $x^2 - 6x + 3 = (x - 3)^2 - 9 + 3$ ✔✔

$= (x - 3)^2 - 6$

Hence the coordinates of the turning point are $(3, -6)$ ✔

Marking

Score 1 mark for the correct value in the brackets and 1 mark for either -9 on line 1 or -6 on line 2.
Both parts of the coordinate point should be correct for the final mark.

4 (a) $y = x - 1$ ✔

(b) $x = 4.2$ or -0.2 ✔

Rearrange the equation to make the left-hand side the same as the given graph.

12 Transformations of graphs

1

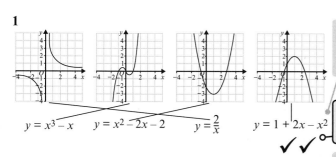

$y = x^3 - x$ $y = x^2 - 2x - 2$ $y = \frac{2}{x}$ $y = 1 + 2x - x^2$

✔✔

Make sure you can recognise the basic shapes of positive and negative quadratic and cubic graphs, and reciprocal graphs.

Marking

Score 1 mark for two or three correct and score both marks if all four are correct.

2 D Exponential ✔

Exponential graphs are continuous curved graphs with no turning points.

3

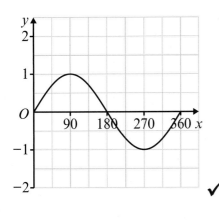

✔ ✔ ○⋯⋯⋯⋯⋯ Learn the shape of the trigonometric graphs so that you can sketch them in an exam.

Marking

Score 1 mark for the correct shape through (0, 0), (180, 0) and (360, 0) and 1 mark for the maximum and minimum points in the correct place.

4

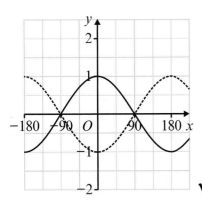

✔ ○⋯⋯⋯⋯ **Core skill**

Graphs of the form $y = -f(x)$ are reflections in the x-axis.

5

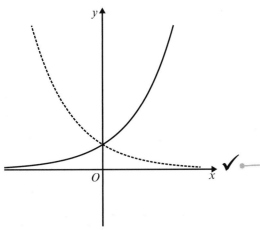

✔ ○⋯⋯⋯ The graph of $y = f(-x)$ is a reflection in the y-axis.

Graphs of the form $y = f(x) + a$ are vertical translations of a units.
Graphs of the form $y = f(x + a)$ are horizontal translations of $-a$ units.

6 To draw the graph of $y = x^2 + 3$ she should translate the graph up by 3 units. ✔ ✔ ○⋯⋯

Marking

Award 1 mark in each case for the correct direction and 1 mark for the correct number of units.

To draw the graph of $y = (x - 2)^2$ she should translate the graph right by 2 units. ✔ ✔ ○

13 Gradients and areas on graphs

1 **A** The car accelerated during the first 20 seconds. ✔ ○⋯⋯

Acceleration is shown by an upwards sloping line, constant speed by a horizontal line and deceleration by a downwards sloping line.
The steeper the line, the faster the rate of acceleration or deceleration.

  C The car travelled at a constant speed between 30 and 50 seconds. ✔

2 **(a)** 46 ✔ ○

Marking

For part (a), score 1 mark if you wrote 45 or 47.
Your answer in this context must be a whole number.

  (b) 16 minutes ✔

3 **(a)**

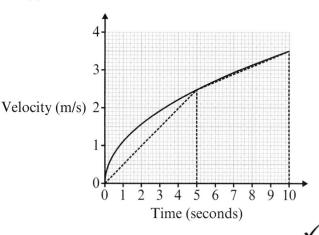

Core skill

Slide your ruler close to the line and choose the angle to get the best tangent.

Try to pick points on the tangent that lie at intersections of the grid. Using (0, 1) and (9, 3.8), the gradient is 2.8 ÷ 9 = 0.311111...

(b) acceleration = $\dfrac{\text{velocity}}{\text{time}} = \dfrac{2.8}{9}$

= 0.31 m/s² ✔ ✔

Marking

For part (b), score 2 marks if your answer is between 0.25 and 0.35 inclusive.

(c)

Velocity (m/s)

Time (seconds) ✔

(d) Area of triangle = 6 units²

Area of trapezium = 15 units² ✔

Area under the graph = 21 units²

Distance travelled = 21 m ✔

Marking

Score 1 mark for one correct area, and both marks for the correct final answer.

(e) **B** Underestimate ✔

The triangle and trapezium do not fill the entire area under the graph.

14 Equation of a circle

The equation of a circle centred on the origin is $x^2 + y^2 = r^2$

1 5 ✔

Remember to halve the diameter to find the radius.

2 **D** $x^2 + y^2 = 36$ ✔

Use Pythagoras' theorem to find the radius. The question asks for the diameter so you must double the radius for the final answer.

3 20 ✔

4 **D** (4, 2) ✔

This coordinate point is the only one that satisfies the equation of the circle.

5 (a) $m_{OP} = \dfrac{1}{3}$ ✔

(b) $m_{\text{tangent}} = -3$ ✔

(c) $y = -3x + 20$ ✔ ✔

6 (a) $m = 4$ ✔

(b) $(1, 4)$ ✔

15 Solving equations

1 $3x - 2 = x + 8$

$2x = 10$

C 5 ✔

2 **C** $x = \dfrac{-b \pm \sqrt{b^2 - 4ac}}{2a}$ ✔

3 $x = -3$ or $x = 2$ ✔

4 **B** $x = 0$ or $x = -5$ ✔

5 $x = \dfrac{-8 \pm \sqrt{8^2 - 4 \times 2 \times (-5)}}{2 \times 2}$

$x = \dfrac{-8 + \sqrt{104}}{4}$ or $x = \dfrac{-8 - \sqrt{104}}{4}$

$x = 0.55$ ✔ or $x = -4.55$ ✔

6 $(x - \boxed{3})(x - \boxed{5}) = 0$ ✔

$x = \boxed{3}$ or $x = \boxed{5}$ ✔

7 (a) $4x^2 - 4x + 1 = x^2 + 5$

$3x^2 - 4x - 4 = 0$ ✔

(b) $(3x + 2)(x - 2) = 0$ ✔

(c) $x = -\dfrac{2}{3}$ or $x = 2$ ✔

8 (a) $x^2 + 4x - 3$

$\equiv (x + \boxed{2})^2 - \boxed{7}$ ✔

(b) **C** $x = -2 \pm \sqrt{7}$ ✔

Core skill

To work out the equation of a tangent to a circle, find the gradient of the tangent and then use the coordinates of the point of contact to find the equation.

Use $y = mx + c$ with $m = -3$ and coordinate $(6, 2)$ to find the equation: $2 = -3 \times 6 + c$ so $c = 20$

Marking

Award 1 mark if you used *your* gradient and the given coordinate point in the equation of the line to try and find c. The 2nd mark is awarded for a complete correct answer.

The gradient of the radius is the negative reciprocal of the gradient of the tangent.

P must have x and y in the ratio $1:4$ and must satisfy the equation of the circle.

Learn the quadratic formula – it won't be given in the exam.

The solutions are the values of x that make each factor equal to 0.

Factorise the left-hand side, so $x(x + 5) = 0$

Core skill

Write in the values you are substituting, then use your calculator to find the answers.

Core skill

Rearrange the quadratic equation so one side is equal to 0. If you can factorise then you can solve the equation without using your calculator.

Marking

Score 1 mark if you got two boxes correct, and both marks if you got all four correct.

Expand the brackets then collect like terms.

Check your answer by expanding the brackets.

Marking

You need both parts correct for the mark.

If $3x + 2 = 0$ then $3x = -2$ so $x = -\dfrac{2}{3}$

$x^2 + bx \equiv \left(x + \dfrac{1}{2}b\right)^2 - \left(\dfrac{1}{2}b\right)^2$

If the left-hand side of a quadratic equation is in completed square form, you can solve it using inverse operations:
$(x + 2)^2 - 7 = 0$
$(x + 2)^2 = 7$
$x + 2 = \pm\sqrt{7}$
Remember to write $\pm$ when you take square roots.

16 Simultaneous equations and iteration

1   **D** $x = 1, y = 2$ ✔

> Substitute each possible pair into both equations to check they work.

2 $4x - 6y = 10$
+
     $9x + 6y = 42$

       $13x = 52$

       $x = 4$

       $8 - 3y = 5$

       $3y = 3$

       $y = 1$

       $x = 4, y = 1$ ✔ ✔

Core skill

> You can solve linear simultaneous equations by elimination. Add or subtract multiples of each equation to eliminate y, then solve to find x.

Marking

> Give yourself 1 mark for each correct value.

3   **D** $x = 5, y = -1$ ✔

> Substitute each possible pair into **both** equations to check they work.

4 $x^2 + (x + 4)^2 = 26$

     $x^2 + x^2 + 8x + 16 = 26$

     $x^2 + 4x - 5 = 0$ ✔

     $x = -5, y = -1$ ✔

     $x = 1, y = 5$ ✔

Core skill

> Substitute for either x or y in the non-linear equation, expand and simplify to form a quadratic equation, which can then be solved.
> If one equation contains a squared term you might need to find **two pairs** of values.

Marking

> Give yourself 1 mark for writing down the correct quadratic equation, and 1 mark for each correct pair of values.

5   **(a)** $2x = x^3 + 1$

       $x = \dfrac{x^3 + 1}{2}$

       $a = 1, b = 2$ ✔

Core skill

> Use inverse operations to rearrange a given function equal to zero into the form $x = \ldots$

   (b) $x_1 = 0.5$

       $x_2 = 0.5625$

       $x_3 = 0.5889\ldots$ ✔

> Use the ANS button on your calculator to speed up this process.

   (c) $x = 0.6$ ✔

> From the answers to part (b) it looks like the value of x tends toward 0.6

17 Using equations and inequalities

1   **B** $x \geqslant 7$ ✔

> $4x \geqslant 28$ therefore $x \geqslant 7$

2   **(a)** $2x + 4y = 31$

       $4x + 5y = 50$ ✔

Marking

> Both equations need to be correct to score 1 mark.

   (b) **C**   Adult: £7.50; Child: £4 ✔

> You can check each pair of values in both equations rather than solving the equations simultaneously.

3

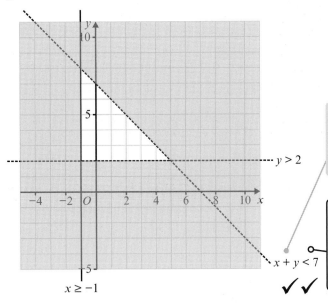

$y > 2$

$x + y < 7$

$x \geqslant -1$

✔ ✔

Use dotted lines for < and > inequalities and solid lines for ⩾ and ⩽.
Choose a point in your region and check that the x- and y- coordinates satisfy the inequalities.

Marking

Score 1 mark for two correct lines (including whether they are dotted or solid).
Score the 2nd mark if all your lines are correct and you have clearly marked your region by shading the area that does not satisfy the inequalities.

4 $2x + x + 10 + 3x - 10 = 180$ ✔

$x = 30$

Angles are: $2 \times 30 = 60$; $30 + 10 = 40$;
$3 \times 30 - 10 = 80$

Largest angle = $\underline{80°}$ ✔

Solve an equation in x then substitute to find the largest angle.

5 **(a)** $x^2 - 6x + 8 = \underline{(x - 4)(x - 2)}$ ✔

(b) $\underline{2 < x < 4}$ ✔

(c)

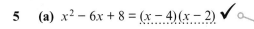

✔

Core skill

Find the critical values and then decide whether the values of x satisfying the inequality are a single 'in between' inequality or two separate 'outside' inequalities.

Use at least one value of x between the two critical values to check.

6 $(2x + 1)(4x - 3) = 78$

$8x^2 - 2x - 81 = 0$ ✔

$x = \underline{3.31}$ ✔

Form a quadratic equation equal to 0 from the information given.
Only the positive solution is required.

18 Sequences

Marking

Give yourself 1 mark for each correct term in the sequence.

1 $\underline{23}$ and $\underline{37}$ ✔ ✔

2 **C** $\underline{25}$ ✔

3 **(a)** $\underline{48}$ ✔

Substitute $n = 9$ into the expression for the nth term.

(b) $\underline{63}$ ✔

You can try a few values for n.

4 **(a) B** $\underline{No}$ ✔

Core skill

Try a few values for n or set up and solve an equation to work out if a given number is part of a sequence.

(b) $u_n = \underline{u_{n-1} + 4}$ ✔

The number in front of n tells you how much you add to the previous term to get the next term.

5 **(a)** The name given to a sequence of this type is a geometric sequence. ✔

(b) $u_1 = 2^1 = 2$; $u_2 = 2^2 = 4$;
$u_3 = 2^3 = 8$ 2, 4 and 8 ✔

6 **D** $3^n - 2$ ✔

7 **A** 4 ✔

19 Finding nth terms

1 **B** A quadratic sequence ✔

2 Zero term

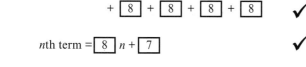

nth term = $\boxed{8}\, n + \boxed{7}$ ✔

3 Difference between terms = $8 - 5 = 3$

Zero term = $5 - 3 = 2$

$3n + 2$ ✔ ✔

4 **C** $13 - 3n$ ✔

5 **D** $n^2 - 1$ ✔

6 $n^2 + 2n + 3$ ✔ ✔

7 $2n^2 + 1$ ✔ ✔

8 $\frac{1}{2}n^2 - 5n + 3$ ✔ ✔ ✔

Ratio and proportion

20 Units, scale drawings and maps

Do each stage of the converson separately. Find the distance travelled in metres in 1 hour first and then express this in terms of kilometres.

1 $10\,\text{m/s} = \underline{10 \times 60 \times 60}\,\text{m/h}$

   $= 36\,000\,\text{m/h}$

   $36\,000\,\text{m/h} = (36\,000 \div 1000)\,\text{km/h} = \underline{36}\,\text{km/h}$ ✔

Marking
Give yourself 1 mark for dividing by 32.

2 $4.8\,\text{m} = (4.8 \times 100)\,\text{cm} = 480\,\text{cm}$

   $480 \div 32$ ✔ $= \underline{15}\,\text{cm}$ ✔

Convert 4.8 m to cm. The actual car is 32 times bigger than the model, so divide by 32.

3   C $\underline{211°}$ ✔

Bearings differ by 180°. It might be helpful to draw a sketch map.

4   **(a)**

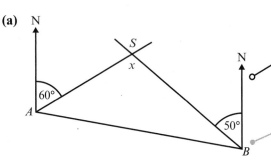

Marking
Give yourself 1 mark for drawing the bearing of S from A correctly, or the bearing of S from B correctly.

Bearings are measured clockwise from north. They always have three figures, so you need to add zeros if the angle is less than 100°. In this example, the bearing of S from A is 060°. The angle at B is 360° − 310° = 50°.

   (b)   Length of $AS = 3.7\,\text{cm}$

   $3.7 \times 500\,000 = 1850\,000\,\text{cm}$ ✔

Marking
Give yourself 1 mark for any length between 3.5 and 3.9 cm or for multiplying your length by 500 000.

   $1850\,000\,\text{cm} = 1850\,000 \div 100\,000 = 18.5\,\text{km}$

   $= \underline{19}\,\text{km (to 2 s.f.)}$ ✔

Marking
If you have used a length between 3.5 cm and 3.9 cm, give yourself 1 mark for 18 km, 19 km or 20 km.

5 $42\ \text{miles} = 42 \div \boxed{5} \times \boxed{8} = \boxed{67.2}\,\text{km}$ ✔

   42 miles per gallon = 67.2 km per gallon
   $= 67.2\,\text{km per 4.55 litres}$

   $67.2 \div 4.55 = 14.769$ ✔

   $14.769 = \underline{14.8}\,\text{km per litre (to 3 s.f.)}$ ✔

Marking
Give yourself 1 mark for two out of three boxes correct.

Marking
Give yourself 1 mark for dividing by 4.55.

5 miles = 8 km, so 1 mile = $\frac{8}{5}$ km. To convert from miles to km divide by 5 and multiply by 8.

21 Ratio

1   **(a)** $\frac{5}{9}$ are red so $\boxed{\dfrac{4}{9}}$ are white

   ratio of red to white is $\boxed{\dfrac{5}{9}} : \boxed{\dfrac{4}{9}} = \underline{5:4}$ ✔

If $\frac{5}{9}$ are red, then $1 - \frac{5}{9} = \frac{4}{9}$ are white and $\frac{5}{9} : \frac{4}{9}$ in its simplest form is 5 : 4

   (b)   Largest possible number of balls = $\underline{999}$ ✔

   largest number of red balls = $\frac{5}{9} \times 999 = 555$

   largest number of white balls = $\frac{4}{9} \times 999 = 444$

   red: $\underline{555}$

   white: $\underline{444}$ ✔

There are whole number of red and white balls in the pool represented by $\frac{5}{9}$ and $\frac{4}{9}$ of the total, so the total must be a multiple of 9. The maximum number is the largest number less than 1000 which is a multiple of 9.

Marking
Give yourself 1 mark for working out the largest number of balls that can be in the ball pool.

2 £25 = $\boxed{2}$ parts

25 ÷ 2 = £12.50

£$\boxed{12.50}$ = 1 part ✔

Total number of parts = 5 + 3 = 8

Total amount of money = 8 × £12.50 ✔

$\qquad\qquad\qquad$ = £100 ✔

Marking
Give yourself 1 mark for both boxes correct.

Matt gets 5 parts of the money and Emma gets 3 parts.

Marking
Give yourself 1 mark for 8 × £12.50

Core skill
Read ratio questions carefully. In this example, Emma gets 3 parts and Matt gets 5 parts, so there are 8 parts in total. £25 = 5 − 3 parts. Work out the value of 1 part then find the total.

3 At the start:

number of white cards = $\boxed{56 ÷ 2 = 28}$

number of black cards = $\boxed{56 ÷ 2 = 28}$ ✔

7 white cards removed:

number of white cards = 28 − 7 = 21

number of black cards = 28

The ratio of white cards to black cards is 21:28 ✔

21:28 = 3:4 ✔

If the ratio is 1:1 then the number of white cards = the number of black cards = 56 ÷ 2 = 28
Make sure you show your working clearly.

Marking
Give yourself 1 mark for both boxes correct.

Marking
Give yourself 1 mark for working out that after 7 white cards are removed, the ratio of the number of white cards to the number of black cards = 21:28.

Marking
Give yourself 1 mark for working out 15:20:4

4

milk	:	dark	:	white
$\boxed{3}$	:	$\boxed{4}$		
		$\boxed{5}$	:	$\boxed{1}$
$\boxed{15}$	:	$\boxed{20}$	:	$\boxed{4}$

15 + 20 + 4 = 39 total parts

4 parts white chocolate ✔

$\frac{80}{39} × 4 = 8.205$ ✔

so greatest number of white chocolates is 8

The two ratios have dark chocolate in common. The LCM of 4 and 5 is 20 so write 3:4 as 15:20 and 5:1 as 20:4 then you can write milk:dark:white as 15:20:4

Marking
Give yourself 1 mark for $\frac{80}{39} × 4$

22 Ratio (continued)

1 (a) 250 g = $\boxed{2}$ parts

$\boxed{125}$ g = 1 part

3 parts = 3 × 125 = 375 g

Total weight = 250 + 375 = 625 g ✔

In this question you are asked to find the total weight so don't forget to add the two weights together.

(b) $\dfrac{F}{B} = \dfrac{\boxed{3}}{\boxed{2}}$ ✔

$F = \dfrac{3}{2}B$ or $F = 1.5B$ ✔

Marking
Give yourself 1 mark for both boxes correct.

Core skill
You are asked for an equation for F in terms of B so it must be of the form $F = \ldots$

2 0.5 litres = $\boxed{500}$ ml

cordial : water = $\boxed{1}$: $\boxed{5}$ ✔

> **Marking**
> Give yourself 1 mark for 3 boxes correct.

3 litres of water needs 3 ÷ 5 = 0.6 litres cordial ✔

> **Marking**
> Give yourself 1 mark for 0.6 *l* or 600 ml of cordial.

3 + 0.6 = 3.6 litres of drink ✔

> You could also work out this question using proportion:
> cordial: 100 ml × 6 = 600 ml
> water: 0.5 litres × 6 = 3 litres
> Total = 600 ml + 3 litres = 3.6 litres

3 $4y = 5x$

$y = \dfrac{5}{4}x$ ✔

$\dfrac{y}{x} = \dfrac{5}{4}$ so $y : x = 5:4$ ✔

> **Marking**
> Give yourself 1 mark for rearranging to get $y = \dfrac{5}{4}x$

> You need to rearrange the equation to find $\dfrac{y}{x}$

4 $\dfrac{2x}{x-2} = \dfrac{\boxed{10x}}{\boxed{2x-1}}$ ✔

Multiplying both sides
by $(x-2)(2x-1)$ gives

$2x(2x-1) = 10x(x-2)$

$4x^2 - 2x = 10x^2 - 20x$ ✔

> **Marking**
> Give yourself 1 mark for getting a correct quadratic.

Rearranging and collecting like terms

$6x^2 - 18x = 0$

$6x(x-3) = 0$ ✔

> **Marking**
> Give yourself 1 mark for factorising correctly.

$x = 0$ and $x = 3$ ✔

23 Percentages

1 $\dfrac{\boxed{36}}{\boxed{48}} \times \boxed{100} = 75\%$ ✔

> To write one quantity as a percentage of another
> • divide the first quantity by the second quantity
> • multiply your answer by 100.

2 multiplier for increase of 8.5% = $\boxed{1.085}$

multiplier for decrease of 12% = $\boxed{0.88}$ ✔

$1.085 \times 0.88 = 0.9548$ ✔

> To find the multiplier for the overall percentage change, multiply the two multipliers together. If the answer is **less than 1** (as here), this shows an overall percentage **decrease**.

3 Actual increase = $\boxed{132} - \boxed{96}$ ✔

$132 - 96 = 36$

$\dfrac{36}{96} \times 100$ ✔

$\dfrac{36}{96} \times 100 = 37.5\%$ ✔

> **Core skill**
> To calculate a percentage change
> • work out the amount of the change
> • write this as a percentage of the original amount.

4 Interest for 1 year = $\dfrac{\boxed{2.4}}{\boxed{100}} \times \boxed{600}$ ✔

$\dfrac{2.4}{100} \times 600 = £14.40$

Interest for 4 years = 4 × Interest for
1 year = 4 × £14.40 ✔

> **Marking**
> Give yourself 1 mark for multiplying the interest for 1 year by 4.

= £57.60 ✔

> To work out simple interest
> • work out the amount of interest for 1 year
> • multiply your answer by the number of years.

5 Using multipliers:

salary $2017 \times 1.025 \times 1.032 = £26\,445$ ✔

$26\,445 \div 1.032 \div 1.025 = $ salary in 2017 ✔

salary in 2017 $= £25\,000$ ✔

This is an application of using reverse percentages.

24 Compound measures

1 Time $= \boxed{530} \div \boxed{75}$ minutes ✔

$530 \div 75 = 7.0666 \dots$ minutes

$0.0666 \times 60 = 4$

7 minutes 4 seconds ✔

Divide 530 by 75 to find the total number of minutes. To convert to seconds, multiply the decimal part of the number by 60 because there are 60 seconds in 1 minute.

2 2 hours 15 minutes $= \boxed{2.25}$ hours ✔

average speed $= \dfrac{\text{distance}}{\text{time}} = \dfrac{63}{2.25} = 28\,\text{km/h}$ ✔ ✔

Core skill
$$\text{average speed} = \frac{\text{distance}}{\text{time}}, \quad \text{time} = \frac{\text{distance}}{\text{average speed}}$$
$$\text{distance} = \text{average speed} \times \text{time}$$

3 (a) $A = \dfrac{F}{p}$ ✔

$A = \dfrac{120}{25} = 4.8\,\text{m}^2$ ✔ ✔

4 Using mass = density × volume:

Mass of orange juice $= \boxed{1.25} \times \boxed{220} = 275\,\text{g}$

Mass of lemon juice $= \boxed{1.03} \times \boxed{180} = 185.4\,\text{g}$

Use the formula triangle for pressure to find the equations for force and area. Always state the units with your answer if they are not given.

Mass of sparkling water $= \boxed{0.99} \times \boxed{750} = 742.5\,\text{g}$ ✔

Total mass $= 275 + 185.4 + 742.5 = 1202.9\,\text{g}$ ✔

Density $= \dfrac{\text{mass}}{\text{volume}} = \dfrac{1202.9}{1150}$ ✔

$\dfrac{1202.9}{1150} = 1.046 = 1.05$ grams per cm^3 to 2 decimal places ✔

25 Proportion

1 D $y = 2x$ ✔

2 C

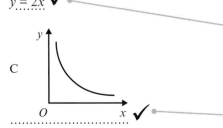

If y is directly proportional to x:
- you can write $y \propto x$
- the graph of y against x is a straight line, passing through the origin
- you can write an equation $y = kx$ where k is a number.

If y is inversely proportional to x, the graph of y against x looks like a reciprocal graph

3 **D** It is divided by 4 ✔

4 1 British pound = 232 ÷ 200 euros ✔

$£1 = €1.16$ ✔

5 1 man would take $\boxed{9} \times \boxed{8}$ days ✔

1 man takes $9 \times 8 = 72$ days

3 men take $72 \div 3$ ✔

$= 24$ days ✔

6 **(a)** $x \propto \dfrac{\boxed{1}}{\boxed{y^2}}$ so $x = \dfrac{\boxed{k}}{\boxed{y^2}}$ ✔

When $y = 6$, $x = 7.5$

Substituting into $x = \dfrac{k}{y^2}$

$7.5 = \dfrac{k}{6^2}$ ✔

$k = 7.5 \times 36 = 270$, so $x = \dfrac{270}{y^2}$ ✔

(b) When $y = 5$, $x = \dfrac{\boxed{270}}{\boxed{5^2}}$

$x = 10.8$ ✔

26 Rates of change

1 **(a)**

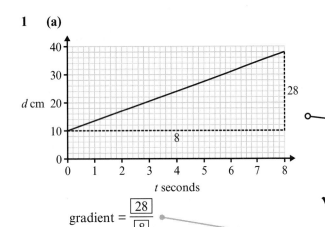

gradient $= \dfrac{\boxed{28}}{\boxed{8}}$

$m = 3.5\,\text{cm/s}$ ✔

(b) Gradient represents the rate of change of depth in cm with time in seconds. ✔

If y is inversely proportional to x:
- you can write $y \propto \dfrac{1}{x}$
- you can write an equation $y = \dfrac{k}{x}$ where k is a number.

Dividing the number of euros by the number of pounds gives the conversion rate from British pounds to euros, so $£1 = 1.16$ euros

Marking
Give yourself 1 mark for 9×8

Marking
Give yourself 1 mark for dividing by 3

At each stage of your working, think, 'should my answer be bigger or smaller?'

Marking
Score 1 mark for either fraction correct.

Marking
Score 1 mark for correctly substituting $y = 6$, $x = 7.5$ into $x = \dfrac{k}{y^2}$

Core skill
To find a proportionality formula for direct and inverse proportion:
- write down the formula using k as the constant of proportionality
- substitute the values you are given
- solve the equation to find the value of k
- write down the formula putting in the value of k.

You can then use your formula to work out the value of one variable if you know the other.

Marking
Give yourself 1 mark for drawing a suitable triangle.

Use points on the graph where the values of the two variables are clear. Be careful reading the different scales on the two axes.
Remember to include the units.

Marking
Give yourself 1 mark for rate of change of depth with time.

2

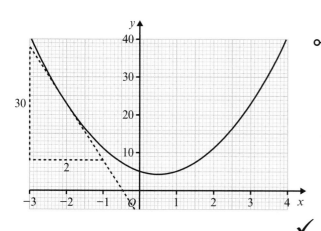

Marking
Give yourself 1 mark for drawing a suitable triangle.

Marking
Give yourself 1 mark for dividing the vertical by the horizontal.

Gradient $= -\dfrac{30}{2}$ ✔

$m_P = \underset{.......}{-15}$ ✔

Core skill
Estimate the gradient of a curve at a point by drawing a tangent to the curve at that point. The gradient represents the rate of change of the variable on the vertical axis with the variable on the horizontal axis. In distance–time graphs, the gradient represents the speed at that point. On a velocity–time graph the gradient represents the acceleration.

3 (a)

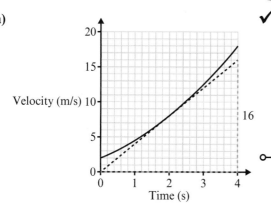

Marking
Give yourself 1 mark for line drawn that has approximately the same gradient as the curve at the point where $t = 2$

gradient $= \underset{........}{4\,\text{m/s}^2}$ ✔

Remember to include the units.

(b) The gradient of a velocity–time graph represents the $\underset{..........}{\text{acceleration}}$. ✔

4

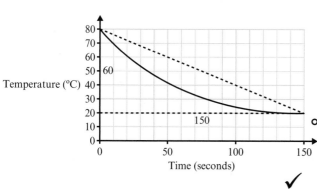

Marking
Give yourself 1 mark for line drawn joining the end points from $t = 0$ to $t = 150$

Average rate of decrease $= \dfrac{60}{150} = \underset{..........}{0.4\ ^\circ\text{C/s}}$ ✔

To work out the average rate of change between $t = 0$ and $t = 150$, draw a straight line between these points on the graph and find its gradient.
This is $\dfrac{\text{change in temperature}}{\text{change in time}}$

27 Growth and decay

1 $A = 500 \times \boxed{1.03}^{\boxed{4}}$ ✔

$A = 500 \times 1.03^4 = 562.754\ldots$

$A = \underline{£562.75}$ ✔ ○──────

> **Core skill**
> Compound interest is an example of percentage change:
> - final amount = starting amount × (multiplier)^n
> where n is the number of times the change is made.
> Remember to write answers to money calculations to the nearest penny unless told otherwise.

2   **(a)** After 3 complete years $n = 3$,

$V = 30\,000 \times 0.9^{\boxed{3}}$ ✔ ○──

$30\,000 \times 0.9^3 = \underline{£21\,870}$ ✔ ●──

> **Marking**
> Give yourself 1 mark for writing 3 in the box.

> The multiplier is less than 1, so the value of the car is decreasing.
> Check that your answer is less than the original value.

   (b) $n = 3$ $V = 21\,870$

   $n = 4$ $V = 19\,683$

   $n = 5$ $V = 17\,714.7$ ○──

   $n = 6$ $V = 15\,943.23$ ✔

   $\underline{6 \text{ years}}$ ✔ ●──────

> **Marking**
> Give yourself 1 mark for writing at least 2 of the values of V correctly.

> Make sure you know how to use your calculator for these types of questions.

3   **(a)** $P = 4200 \times 1.04^n$, where n is the number of years after 2010.

   In 2010, $n = 0$, so $P = 4200 \times 1.04^0$

   $1.04^0 = 1$ so $\underline{P = 4200}$ ✔ ●──

> n represents the number of years after 2010, so in 2010 $n = 0$

   (b) $n = 2025 - 2010 = \boxed{15}$ ✔ ○──

   $P = 4200 \times 1.04^{15} = 7563.96$ so estimate for population is $\underline{7563}$ ✔ ●──

> **Marking**
> Give yourself 1 mark for $n = 15$

> Your answer is an estimate of the number of people, so it must be a whole number.
> In this type of question, you always round down your answer to the next integer below.

4   **(a)** $x^3 + 5x - 2 = 0$

   Let $\underline{f(x) = x^3 + 5x - 2}$

   $f(0) = \underline{-2}$

   $f(1) = \underline{1 + 5 - 2 = 4}$ ✔

   $\underline{\text{Change of sign indicates a root between}}$ ○──
   $\underline{0 \text{ and } 1}$ ✔

> **Core skill**
> If there is a solution between two values, it means there is a root of the equation. This means that the curve crosses the x axis so the sign of $f(x)$ will change. Note that you may first have to rearrange the equation so that $f(x) = 0$

   (b) $x_{n+1} = \dfrac{2}{5} - \dfrac{x_n^3}{5}$ and $x_0 = 0$

   For $n = 0$:

   $x_1 = \dfrac{2}{5} - \dfrac{x_0^3}{5} = \dfrac{2}{5}$

   $x_1 = 0.4$ ✔ ○──────

> **Marking**
> Give yourself 1 mark for $x_1 = 0.4$

   For $n = 1$:

   $x_2 = \dfrac{2}{5} - \dfrac{x_1^3}{5} = \dfrac{2}{5} - \dfrac{(0.4)^3}{5}$

   $x_2 = \underline{0.3872}$ ✔ ○──

> **Core skill**
> In an iterative process, you are given a starting value and each consecutive term is worked out from the previous term. The more iterations you do, the closer you get to the value of the root. If you are asked to give the answer correct to 3 significant figures, you need to repeat the process until two successive terms agree to 3 significant figures.

Geometry and measures

28 Angle problems

1 **(a)** $x = \underline{105°}$ ✔

Core skill
It is important that you use the correct terminology when giving reasons.

Reason: <u>Angles on a straight line add to 180°</u> ✔

Marking
1 mark for the correct angle, 1 mark for the correct reason.

(b) $y = \underline{75°}$ ✔

Reason: <u>Corresponding angles are equal</u> ✔

2 **(a)** $\underline{45°}$ ✔

Size of one exterior angle = 360° ÷ number of sides.

(b) $\underline{135°}$ ✔

180° − 45° = 135°

3 $\underline{17}$ ✔

Sum of interior angles = $(n - 2) \times 180°$

4 $(2x - 10) + (3x + 20) + (4x - 30) = 180$ ✔

$x = \dfrac{200}{9} = 22.22°$

Largest angle = $\underline{86.67°}$ ✔

Marking
1 mark for forming a correct equation, 1 mark for finding the largest angle.

Read the question carefully. You are asked to find the largest angle, so you must substitute the value of x into the three expressions to find the angle.

5 **(a)** $x = \underline{49°}$ ✔

(b) Angle opposite 78° = 78°, angle on same line as <u>127° = 53°. Interior angles in a triangle are 180° so $x = 180 − 53 − 78 = 49°$</u> ✔

Use the diagram to help explain your reasoning. Mark angles you have calculated onto the diagram and make sure that you use the correct terminology when you write your answer.

29 Constructions

1

Marking
Score 1 mark for one correct circle (a 5 cm radius circle drawn centred on A, or a 4 cm radius circle centred on B) and the 2nd mark for both circles correct and the correct shading in the intersection of the two circles.

2

Core skill
Use standard ruler and pair of compass constructions to solve problems involving loci.
1 mark for each correct locus drawn and 1 mark for shading the correct region.

3

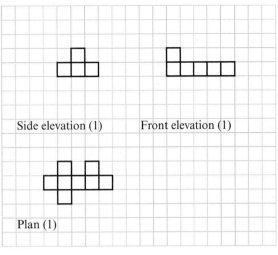

Side elevation (1) Front elevation (1)

Plan (1)

✔ ✔ ✔

4 The 3D shape has 8 faces, 18 edges and 12 vertices. ✔ ✔ ✔

> The 3-D shape is a hexagonal prism.

30 Congruence and similarity

1 D SSA ✔

> The angle must be between the two sides.

2 (a) ⬚5⬚ : ⬚2⬚ ✔

(b) 15 cm ✔

> **Core skill**
> Use the lengths of the corresponding sides to work out the scale factor of enlargement between the two triangles.

3 (a) D ✔

(b) 9 cm ✔

> The scale factor of enlargement is $6 \div 8 = 0.75$
> $12 \times 0.75 = 9$

4 1: Both triangles are right-angled ✔

2: AC is common to both triangles ✔

3: $BC = CD$ ✔

4: Right-angle, hypotenuse and a side are equal (RHS) ✔

> **Marking**
> You can give the facts in any order. State the congruence criterion you have demonstrated to get the final mark.

5 Ratio of corresponding sides: $\frac{x}{3}$ or $\frac{3}{x}$ ✔

$\frac{3}{x} = \frac{x}{BC}$

$BC = \frac{x^2}{3}$ cm ✔

> **Marking**
> Award 1 mark for the correct ratio of sides and 1 mark for writing BC in terms of x.

> Make sure you get the ratios the right way up when setting up the equation.

31 Transformations

1 (a) Rotation ✔

180° about (0, 0) ✔

(b) Reflection ✔

In the line $x = -1$ ✔

> **Core skill**
> Make sure that you give all of the information required to describe fully a single transformation that maps one shape onto another.

> **Marking**
> Score 1 mark for rotation and 1 mark for the angle and centre of rotation.

> **Marking**
> Score 1 mark for reflection and 1 mark for the line of reflection.

2

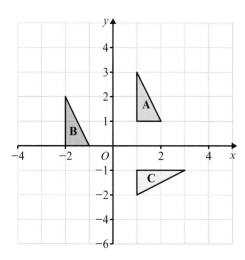

Use tracing paper to help with the rotation.

Marking

Score 1 mark for each shape you transformed correctly.

✓ ✓

3

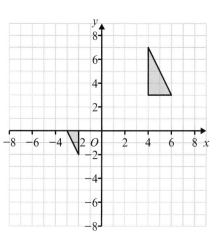

Score 1 mark for the correct size of your image and 1 mark for the correct position and orientation.

Draw the two transformations successively on the diagram and identify any points on the image that are in the same place as in the original triangle.

✓ ✓

4 (2, 1) ✓ ✓

Marking

Score 2 marks for a correct coordinate pair. Score 1 mark if you have given the correct answer but also included other coordinates.

32 Circle facts and theorems

1 **C** Angles in the same segment add up to 180° ✓ · · · Angles in the same segment are equal.

2 $x = 63°$ ✓

Core skill

Give reasons in all cases.

Reason: Angle at the centre is twice the angle at

the circumference. ✓

3 $x = 78°$ ✓

Angle $CBT = 37°$ (Alternate segment theorem) ✓

Angle $ABC = 102°$ (Angles on a straight line) ✓

Marking

Score 1 mark for each reason, up to a maximum of three, and score the final mark for the correct value of x.

Angle $ADC = 78°$ (Opposite angles in a cyclic

quadrilateral add up to 180°) ✓

There are different ways of working out x, which use different reasons. You can score the marks for any method as long as you give your reasons.

4 **(a)** Angle $OCA = x$ Angle $OCB = y$ ✓

OA, OB and OC are radii, so $OA = OB = OC$

Reason: Base angles in isosceles triangle. ✓

(b) $x + x + y + y = 180°$ ✓
Angles in a triangle add up to 180° ✓
$2x + 2y = 180°$
$x + y = 90°$
So angle $ACB = 90°$ ✓

Marking

Give yourself 1 mark for **both** correct angles, and 1 mark for the correct reason.

Use the fact that the angles in a triangle add up to 180°. Remember to state any angle facts you use in your working.

33 Area and perimeter

1 $27\,\text{cm}^2$ ✓

> Area of a triangle $= \frac{1}{2} \times \text{base} \times \text{height} = \frac{1}{2} \times 9 \times 6$

2 $18\,\text{cm}$ ✓

> Work backwards from the formula for the area of the circle to find r. Don't forget that the question asks you for the diameter.

3 (a) $\dfrac{\boxed{110}}{\boxed{360}} \times \pi \times \boxed{7}^{\boxed{2}} = 47.0\,\text{cm}^2$ ✓

> **Core skill**
> Sector areas and arc lengths are fractions of the area and circumference of the circle respectively.

  (b) Arc length $= 13.4\dots\,\text{cm}$ ✓

  Perimeter $= 13.4 + 7 + 7 = 27.4\,\text{cm}$ ✓

> Add on the two radii to calculate the perimeter of the sector.

4 **B** $100\,000\,\text{cm}^2$ ✓

> $1\,\text{m}^2 = 100 \times 100 = 10\,000\,\text{m}^2$, so $10\,\text{m}^2 = 100\,000\,\text{m}^2$

5 $9 \times 15\,000 = 135\,000\,\text{cm}$

  $1.35\,\text{km}$ ✓

> Divide by 100 to find the distance in m and by 1000 to find the distance in km.

6 $430.5 + 357$ ✓

  $= 787.5\,\text{cm}^2$ ✓

> Area of a trapezium $= \frac{1}{2}(a + b)h = \frac{1}{2}(24 + 17) \times 21$
>
> Area of parallelogram $= bh = 21 \times 17$

7 $\dfrac{3}{4} \times 2 \times \pi \times 3.5 + \dfrac{3}{4} \times 2 \times \pi \times 3$

  $+ 2 \times 0.5 = 31.63\dots\,\text{m}$ ✓

  $31.63 \times 11.50 = \pounds363.75$ ✓

> **Marking**
> Award yourself 1 mark for 430.7 or 357. Award yourself the second mark if you got both areas correct and the correct final answer.

> **Marking**
> Score 1 mark for attempting to find the two arc lengths.

34 Surface area and volume

1 Area of cross-section $= \dfrac{5 + 8}{2} \times 6 = 39\,\text{cm}^2$ ✓

  Volume $= 702\,\text{cm}^3$ ✓

> **Core skill**
> Calculate the cross-section area first, then multiply this by the length.

> **Marking**
> Score 1 mark for the correct cross-section area and 1 mark for the correct volume.

2 $180\pi \div (1.5^2 \times \pi)$ ✓

  $= 80\,\text{cm}$ ✓

> Divide the volume of the cylinder by the formula for the cross-section area.

> **Marking**
> Score 1 mark for dividing the volume by at least π or 1.5^2. The 2nd mark is for a complete correct answer.

3 $\dfrac{30}{18} = \dfrac{x + 5}{x} \Rightarrow 5x = 3x + 15$ ✓

  $x = 7.5$ ✓

> **Marking**
> Score 1 mark for the correct ratio of side lengths. It doesn't have to be written as an equation, or simplified.

4 (a) 320×1.5^2

  $720\,\text{cm}^2$ ✓

  (b) $1440 \div 1.5^3$

  $427\,\text{cm}^3$ (3 s.f.) ✓

> **Core skill**
> Square the ratio of lengths to find the ratio of areas. Cube the ratio of lengths to find the ratio of volumes.

> **Marking**
> In part (b), you could give your answer to any sensible degree of accuracy or leave it exact.

5 $V_{hemisphere} = \frac{4}{3} \times \pi \times 9^3 \div 2 = 486\pi \text{ cm}^3$ ✔

$V_{cone} = \frac{1}{3} \times \pi \times 9^2 \times 14 = 378\pi \text{ cm}^3$ ✔

$486\pi + 378\pi = \underline{864\pi} \text{ cm}^3$ ✔

> The question asks for the **exact** volume of the toy so you must leave your final answer as a multiple of π.

35 Right-angled triangles

> **Core skill**
> Use Pythagoras' theorem to calculate missing side lengths in right-angled triangles.

1 $x = \underline{8} \text{ cm}$ ✔

> You are finding a shorter side so $c^2 - a^2 = b^2$

> **Core skill**
> Use trigonometry and remember SOHCAHTOA.

2 $x = \underline{38.2}°$ ✔

> You have the adjacent side and the hypotenuse so use cosine.

3 Base of small triangle: $7^2 - 5^2 = 24$ so $\sqrt{24} = 4.89...$ ✔

Base of large triangle: $\frac{5}{\tan 21°} = 13.03...$ ✔

$x = 4.89... + 13.03... = \underline{17.9} \text{ cm}$ (3 s.f.) ✔

> Use Pythagoras' theorem to work out the base of the left-hand triangle and trigonometry to work out the base of the right-hand triangle.

> **Marking**
> Score 1 mark for 4.89…, 1 mark for 13.03… and 1 mark for the correct final answer.

4 Side opposite y:

$x = \frac{10}{\cos 40°} = 13.05...$ ✔

$y = \sin^{-1}\left(\frac{13.05...}{25}\right) = 31.47...$

$y = \underline{31.5}°$ ✔

> You need to find the side length opposite y first using the right-hand triangle.

5 Base of tower to L: $\frac{32}{\tan 46°} = 30.90...$

Base of tower to K: $\frac{32}{\tan 35°} = 45.70...$ ✔

$45.7 - 30.9 = \underline{14.8} \text{ m}$ ✔

> **Marking**
> Score 1 mark for using tan correctly at least once.

6 **(a)** $\underline{7.81} \text{ cm}$ ✔

(b) $\sin^{-1}\left(\frac{4}{\sqrt{61}}\right)$ ✔

$= \underline{30.8}°$ ✔

> Use Pythagoras' theorem in 3-D: $a^2 + b^2 + c^2 = d^2$. You could leave your answer in surd form as $\sqrt{61}$.

> **Marking**
> Score 1 mark for a correct trigonometric ratio using your side lengths.

36 Trigonometry

1 **B** $\underline{\sin 60° \text{ and } \cos 30°}$ ✔

2 Base angle = 45°, shorter side lengths are x ✔

$\tan\theta = \frac{\text{opposite}}{\text{adjacent}} = \frac{x}{x} = 1$ therefore $\tan 45° = 1$ ✔

3 **C** $\underline{a^2 = b^2 + c^2 - 2bc\cos A}$ ✔

4 $\frac{x}{\sin 71°} = \frac{3}{\sin 36°}$ ✔

$x = \underline{4.83} \text{ cm}$ ✔

> **Marking**
> Score 1 mark for writing a base angle and indicating the two shorter sides are x. These can be shown on the diagram.
> Score 1 mark for relating these to the tan ratio for an angle of 45° and a complete argument.

> Learn the sine and cosine rules since they will not be given in the exam.

> **Core skill**
> $\frac{a}{\sin A} = \frac{b}{\sin B} = \frac{c}{\sin C}$ or $\frac{\sin A}{a} = \frac{\sin B}{b} = \frac{\sin C}{c}$

> **Marking**
> Score 1 mark for the correct choice of sine rule and the correct values substituted in.

5 $\cos x = \dfrac{4^2 + 11^2 - 9^2}{2 \times 4 \times 11}$ ✔

$\cos x = 0.6363...$ ✔

$x = \underline{50.5°}$ ✔

Remember to find $\cos^{-1}$ at the end. Stopping at $\cos A = ...$ is a common error.

6 Area of sector: $\dfrac{80}{360} \times \pi \times 5^2 = 17.45...$ ✔

Area of triangle: $\dfrac{1}{2} \times 5^2 \times \sin 80° = 12.31...$ ✔

Area of segment $= \underline{5.14}\,\text{cm}^2$ (3 s.f.) ✔

Calculate the area of the sector and then subtract the area of the triangle. $A = \frac{1}{2}ab \sin C$

37 Vectors

1 $\begin{pmatrix} 2 \\ -1 \end{pmatrix}$ ✔

The shape moves two spaces to the right and one space down.

2 **(a)** $\begin{pmatrix} 7 \\ 1 \end{pmatrix}$ ✔

(b) $\begin{pmatrix} -5 \\ 7 \end{pmatrix}$ ✔

(c) $\begin{pmatrix} 27 \\ -1 \end{pmatrix}$ ✔ ✔

3 $2\mathbf{a} + \begin{pmatrix} 24 \\ -9 \end{pmatrix} = \begin{pmatrix} -8 \\ 20 \end{pmatrix}$

$2\mathbf{a} = \begin{pmatrix} -32 \\ 29 \end{pmatrix}$ ✔

$\mathbf{a} = \begin{pmatrix} -16 \\ 14.5 \end{pmatrix}$ ✔

Solve this problem like you would solve an equation.

4 $\overrightarrow{DB} = \mathbf{a} - \mathbf{b}$ ✔

$\overrightarrow{NM} = \dfrac{1}{2}\mathbf{a} - \dfrac{1}{2}\mathbf{b} = \dfrac{1}{2}(\mathbf{a} - \mathbf{b})$ ✔

Hence $\overrightarrow{NM} = \dfrac{1}{2}\overrightarrow{DB}$ and therefore they are parallel ✔

Probability and statistics

38 Probabilities and outcomes

You do not need to cancel fractions in probability questions. Award yourself marks for any fraction equivalent to the given answer.

1 $P(\text{blue}) = \dfrac{\boxed{4}}{5 + 4 + 3}$ or $\dfrac{1}{3}$ ✔

The probability that the ball is blue is the same as the proportion of blue balls in the bag.

2 **(a)**

		Set 1				
		1	**3**	**5**	**7**	**9**
Set 2	**2**	3	5	7	9	11
	4	5	7	9	11	⑬
	6	7	9	11	⑬	⑮
	8	9	11	⑬	⑮	⑰

This is an example of a sample space diagram that shows you all the possible outcomes of two events. As there are 5 numbers in Set 1 and 4 in Set 2, the total number of possible outcomes is $5 \times 4 = 20$

✔

(b) number of numbers greater than 12 = 6

$$P(\text{greater than } 12) = \frac{6}{20} ✔ ✔$$

3 **(a)** Estimated probability $= \dfrac{48}{220}$ ✔ ✔

(b) $\dfrac{1}{6}$ ✔

(c) Caz's results will give the best estimate for the theoretical probability.

Because they show the highest/largest number of trials. ✔

The theoretical probability is based on the number of times you would expect the outcome to happen.

In the case of a fair dice, you would expect to get a 6 once in every six throws, so the probability is $\dfrac{1}{6}$

4 **(a)** $0.14 + 0.08 + 0.16 + 0.11 + x + 2x = 1$ ✔

$$0.49 + 3x = 1$$

$$3x = 0.51$$

$$x = 0.17 ✔$$

Probability estimates based on relative frequency are more accurate for a larger number of trials in an experiment.

$$p(6) = 2 \times 0.17 = 0.34 ✔$$

The events are mutually exclusive (they cannot happen at the same time) so the probabilities add up to 1.

(b) $200 \times 0.16 = 32$ ✔

Expected number of outcomes = number of trials × probability

39 Venn diagrams

1 x = number of students who have a brother or a sister.

Total number of students who have a brother or a sister = $60 - 7 = 53$

$$40 - x + x + 35 - x = 53$$

so $75 - x = 53$

$$x = 22 ✔$$

Substitute x back into your original expressions for students with brothers and sisters.

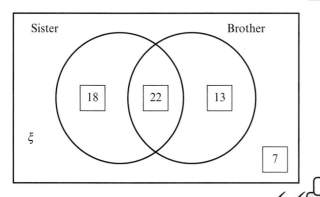

✔ ✔

The symbol ξ is used for the **universal set**. It represents all the elements you have to consider in a question, which in this case is the 60 students. The number in the intersection of the two circles is the number of students that have a brother and a sister.

2 (a) 11 ✔

> You need to count how many numbers appear in the Venn diagram.

(b) (i) $A \cup B = $ 3, 5, 8, 10, 12, 22, 24 ✔

> **Core skill**
> You should know that the symbol ∪ means **union**. The union of two sets is the set of elements that belong to either set.

(b) (ii) $A \cap B = $ 3, 5, 8 ✔

> **Core skill**
> You should know that the symbol ∩ means **intersection**. The intersection of two sets is the set of elements that belong to both sets, in this example it is all the numbers that are in A **and** B.

(c) P(number is in set A') $= \dfrac{7}{11}$ ✔ ✔

> **Marking**
> Give yourself 1 mark if you got one box correct.

> **Core skill**
> You should know that the symbol A' means **not** A or A complement. It is everything in ξ that is not in A. In this example that is 10, 12, 22, 9, 23, 18, 11

3 (a)

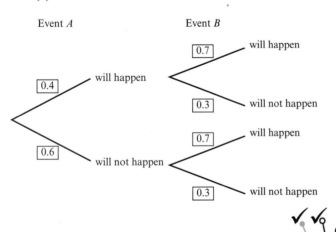

> **Marking**
> Give yourself 1 mark for 4 boxes correct.

> Be careful when filling in the elements on the Venn diagram. There is only one number that is in A and B and C, so this is placed in the overlap of the three circles.

(b) P(a member of $A \cap B'$) $= \dfrac{1}{4}$ ✔ ✔

> **Marking**
> Give yourself 1 mark if either the numerator or the denominator is correct.

> $A \cap B'$ means the element is in A and in everything outside B. Only 9, 13 and 17 satisfy this requirement.

40 Independent events and tree diagrams

1 (a)

Event A Event B

0.4 — will happen
— 0.7 → will happen
— 0.3 → will not happen

0.6 — will not happen
— 0.7 → will happen
— 0.3 → will not happen

> **Marking**
> Give yourself 1 mark for four out of six boxes correct.

> The probabilities on each pair of branches must add up to 1.

(b) P(A will happen and B will not happen) =

$\boxed{0.4} \times \boxed{0.3} = \underline{0.12}$

P(A will not happen and B will happen) =

$\boxed{0.6} \times \boxed{0.7} = \underline{0.42}$ ✔

$0.12 + 0.42 = \underline{0.54}$ ✔

For independent events P(A AND B) = P(A) × P(B)

For tree diagrams:
- multiply along the branches
- add up the outcomes.

For mutually exclusive events P(A OR B) = P(A) + P(B)

2 (a) Let P (Matt wins) = x, then P(Abi wins) = $3x$

and $x + 3x + 0.2 = 1$ ✔

$4x = 0.8$ so P(Matt wins)= $\underline{0.2}$ ✔

(b) P (Matt wins at least 1 of 3 games) = $1 - $ P (Matt wins no games)

P (Matt wins a game) = 0.2, so P (Matt does not win a game) = 0.8

P (Matt wins no games) = $0.8 \times 0.8 \times 0.8$ ✔

$= 0.512$

Remember the sum of the probabilities of all possible outcomes = 1

P (Matt wins at least 1 of 3 games) =

$1 - 0.512 = \underline{0.488}$ ✔

3 P (takes 1 throw) = $\dfrac{\boxed{1}}{\boxed{6}}$

P (takes 2 throws) = $\dfrac{\boxed{5}}{\boxed{6}} \times \dfrac{\boxed{1}}{\boxed{6}} = \dfrac{5}{36}$ ✔

P (takes 3 throws) = $\dfrac{5}{6} \times \dfrac{5}{6} \times \dfrac{1}{6} = \dfrac{25}{216}$

P (takes 4 or more throws) = $1 - $ P (takes 3 or less throws)

P (lands on 6) = $\dfrac{1}{6}$ so P (does not land on 6) = $\dfrac{5}{6}$

$= 1 - [$P (takes 1 throw) + P (takes 2 throws) +

P (takes 3 throws)$]$ ✔

$= 1 - \left(\dfrac{1}{6} + \dfrac{5}{36} + \dfrac{25}{216} \right)$ ✔

$= 1 - \dfrac{91}{216} = \dfrac{125}{216}$ ✔

41 Conditional probability

1 (a)

Event A Event B

```
                                          0.1    sleeps through
                              goes to bed        alarm
                 0.2          before midnight
                                          0.9    does not sleep
                                                 through alarm
                                          0.7    sleeps through
                              goes to bed at     alarm
                 0.8          midnight or later
                                          0.3    does not sleep
                                                 through alarm
```

> This is an example of conditional probability: if one event has already occurred, the probability of the second event changes. You can use a tree diagram to answer questions involving conditional probability.

(b) P(goes to bed before 12 and sleeps through the alarm) = $0.2 \times 0.1 = 0.02$

Marking
Give yourself 1 mark for four out of six boxes correct.

P(goes to bed at 12 or later and sleeps through the alarm) = $0.8 \times 0.7 = 0.56$

Marking
Give yourself 1 mark for one probability correct.

P(Adam sleeps through his alarm on a school night) = $0.02 + 0.56 = 0.58$

2 (a) Number of students that went to Edinburgh = $18 + 16 = 34$

Marking
Give yourself 1 mark for the correct number of students that went to Edinburgh.

P(female | went to Edinburgh) = $\dfrac{16}{34}$

Core skill
You know the student went to Edinburgh, so you only consider the 18 males and 16 females who went there. Of these 34 students, 16 are female.

(b) Number of male students = $15 + 18 + 12 = 45$

P(went to Cardiff | male) = $\dfrac{12}{45}$

Core skill
You know the student was male, so you only consider the 45 male students. Of these 45 students, 12 went to Cardiff.

3 (a) Number of students who play tennis = $12 + 21 + 10 + 9 = 52$

Number of students who go swimming and play tennis = $12 + 10 = 22$

P(swimming | tennis) = $\dfrac{22}{52}$

> Use the Venn diagram to identify the total number of students that play tennis, then identify which of these students also go swimming.

(b) Number of students who take part in at least two activities = $12 + 10 + 8 + 9 = 39$

Number of students who play squash and take part in at least two activities = $8 + 10 + 9 = 27$

P(squash | at least two activities) = $\dfrac{27}{39}$

> In this example you are told the student takes part in at least 2 activities. This means 4 regions on the Venn diagram. Of these 4 regions, 3 lie within the region representing squash. You may find it helpful to shade the restricted sample space on the diagram to help you identify the values you need to work with.

42 Sampling, averages and range

1

Mass (x kg)	Frequency	Midpoint × frequency
$3 \leqslant x < 4$	21	$3.5 \times 21 = 73.5$
$4 \leqslant x < 5$	17	$4.5 \times 17 = 76.5$
$5 \leqslant x < 6$	7	$5.5 \times 7 = 38.5$
$6 \leqslant x < 7$	5	$6.5 \times 5 = 32.5$

Core skill

Use the midpoint of each class and multiply by the frequency. Divide the sum of these values by the total frequency.

✓

$221 \div 50 = \underline{4.42}\,\text{kg}$ ✓ ✓

You could add an extra column onto your table. Show your final answer to an appropriate number of significant figures.

Marking

Score 1 mark for using the midpoint of each class and multiplying by the frequency. Score 1 mark for dividing your total by the total frequency.

2 (a) Median: $\underline{25}$ ✓

$46 - 9$

Range: $\underline{37}$ ✓

There are 25 values so the median is the 13th.

Q_1 is the $(n + 1)/4 = 6.5$th value and Q_3 is the $3(n + 1)/4 = 19.5$th value (halfway between the 19th and 20th values).

(b) $37.5 - 19 = \underline{18.5}$ ✓ ✓

(c) The people on the bus are on average older.

There is a greater spread of ages of people on the bus. ✓

Marking

Score 1 mark for writing 37.5 or 19

Compare both the average and the spread.

3 (a) 1 Ambika could increase the number of people she asks.

2 She could ask people from different classes and/ or years. ✓

Comment on the sample size and comment about who she asks.

(b) She could use an ordered list, e.g. the school register, and generate random numbers. ✓

You could also have said 'put the names in a hat'.

4  $N = \dfrac{50 \times 40}{16}$ ✓

$= \underline{125}$ ✓

$N = \dfrac{Mn}{m}$ where N is the population size, M is the number originally captured and tagged, n is the size of the recapture sample and m is the number tagged in the recapture sample.

43 Representing data

1 (a) Median: $\underline{10.7}\,\text{cm}$ ✓

(b) LQ = 8.8; UQ = 12.6 ✓

Interquartile range = $\underline{3.8}$ ✓

Draw lines on the graph across from the frequency axis to the curve and then read off from the x-axis.

Marking

Score 1 mark for the median if your answer is between 10.6 and 10.8.
Score 1 mark for the LQ and UQ if your answers are in the range 8.7 – 8.9 and 12.5 – 12.7.
The mark for the IQR follows on from *your* values for the LQ and UQ as long as they are in the given ranges.

(c)

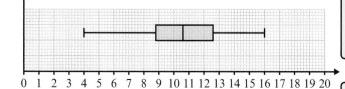

0 1 2 3 4 5 6 7 8 9 10 11 12 13 14 15 16 17 18 19 20

✓ ✓

Marking

Score 1 mark for correctly drawing the box using the median, LQ and UQ, and score the 2nd mark for the correct whiskers.

2 **(a), (b)**

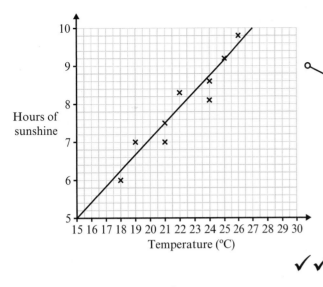

Marking

Make sure your line of best fit has an approximately equal number of points either side.

✔ ✔

(c) (Strong) positive ✔

Marking

You will get a mark for part (d) if you read correctly from *your* line of best fit.

(d) 7.1 hours ✔

(e) Yes: The value is within the range of the data (interpolation). ✔

Core skill

You can make predictions from your line of best fit if the prediction is within the range of the data.

3 **1** 11 000 is outside the range of the data (extrapolation). ✔

You must give a clear reason.

2 Correlation does not imply causation. Both variables are probably linked to a 3rd variable. ✔

You need to learn these reasons.

44 Representing data (continued)

1 **(a)**

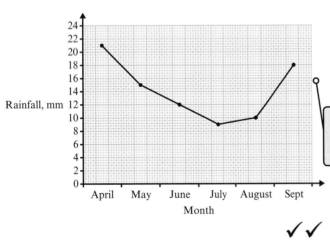

Marking

Score 1 mark for plotting the points correctly and 1 mark for correctly labelled axes and straight line segments drawn between the points.

✔ ✔

(b) The monthly rainfall decreases to July and then increases again. ✔

Make sure you give a clear statement in context.

2 **(a), (b)** Frequency densities: 2, 8.5, 18, 15, 2 ✔

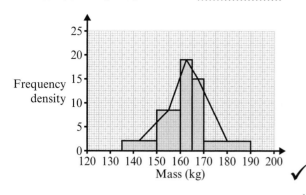

Score 1 mark for calculating frequency densities.
Score 1 mark for a complete correct histogram.

✔ ✔

✔ ✔

Score 1 mark for joining any two adjacent midpoints, and 2 marks for a complete correct frequency polygon.

3 **(a)** The data is in unequal class intervals. ✔

(b) $5 \times 5 + 6 \times 5 + 0.5 \times 20 = 65$ people ✔

Use frequency density × class width to find the frequencies in each complete bar.
Split the last bar into two.